Acute Exposure Guideline Levels for Selected Airborne Chemicals

VOLUME 4

Subcommittee on Acute Exposure Guideline Levels
Committee on Toxicology
Board on Environmental Studies and Toxicology
Division on Earth and Life Studies

NATIONAL RESEARCH COUNCIL
OF THE NATIONAL ACADEMIES

THE NATIONAL ACADEMIES PRESS
Washington, D.C.
www.nap.edu

THE NATIONAL ACADEMIES PRESS 500 Fifth Street, NW Washington, DC 20001

NOTICE: The project that is the subject of this report was approved by the Governing Board of the National Research Council, whose members are drawn from the councils of the National Academy of Sciences, the National Academy of Engineering, and the Institute of Medicine. The members of the committee responsible for the report were chosen for their special competences and with regard for appropriate balance.

This project was supported by Contract Nos. DAMD17-89-C-9086 and DAMD17-99-C-9049 between the National Academy of Sciences and the U.S. Army. Any opinions, findings, conclusions, or recommendations expressed in this publication are those of the author(s) and do not necessarily reflect the view of the organizations or agencies that provided support for this project.

International Standard Book Number 0-309-09147-0 (Book)
International Standard Book Number 0-309-53013-X (PDF)

Additional copies of this report are available from:

The National Academies Press
500 Fifth Street., NW
Box 285
Washington, DC 20055

800-624-6242
202-334-3313 (in the Washington metropolitan area)
http://www.nap.edu

Printed in the United States of America

THE NATIONAL ACADEMIES

Advisers to the Nation on Science, Engineering, and Medicine

The **National Academy of Sciences** is a private, nonprofit, self-perpetuating society of distinguished scholars engaged in scientific and engineering research, dedicated to the furtherance of science and technology and to their use for the general welfare. Upon the authority of the charter granted to it by the Congress in 1863, the Academy has a mandate that requires it to advise the federal government on scientific and technical matters. Dr. Bruce M. Alberts is president of the National Academy of Sciences.

The **National Academy of Engineering** was established in 1964, under the charter of the National Academy of Sciences, as a parallel organization of outstanding engineers. It is autonomous in its administration and in the selection of its members, sharing with the National Academy of Sciences the responsibility for advising the federal government. The National Academy of Engineering also sponsors engineering programs aimed at meeting national needs, encourages education and research, and recognizes the superior achievements of engineers. Dr. Wm. A. Wulf is president of the National Academy of Engineering.

The **Institute of Medicine** was established in 1970 by the National Academy of Sciences to secure the services of eminent members of appropriate professions in the examination of policy matters pertaining to the health of the public. The Institute acts under the responsibility given to the National Academy of Sciences by its congressional charter to be an adviser to the federal government and, upon its own initiative, to identify issues of medical care, research, and education. Dr. Harvey V. Fineberg is president of the Institute of Medicine.

The **National Research Council** was organized by the National Academy of Sciences in 1916 to associate the broad community of science and technology with the Academy's purposes of furthering knowledge and advising the federal government. Functioning in accordance with general policies determined by the Academy, the Council has become the principal operating agency of both the National Academy of Sciences and the National Academy of Engineering in providing services to the government, the public, and the scientific and engineering communities. The Council is administered jointly by both Academies and the Institute of Medicine. Dr. Bruce M. Alberts and Dr. Wm. A. Wulf are chair and vice chair, respectively, of the National Research Council

www.national-academies.org

SUBCOMMITTEE ON ACUTE EXPOSURE GUIDELINE LEVELS

Members

DANIEL KREWSKI *(Chair)*, University of Ottawa, Ottawa, Ontario, Canada
EDWARD C. BISHOP, Parsons Corporation, Pasadena, CA
JAMES V. BRUCKNER, University of Georgia, Athens
DAVID P. KELLY, Dupont Company, Newark, DE
KANNAN KRISHNAN, University of Montreal, Montreal, Quebec, Canada
STEPHEN U. LESTER, Center for Health, Environment and Justice, Falls Church, VA
JUDITH MACGREGOR, Toxicology Consulting Services, Arnold, MD
PATRICIA MCGINNIS, Syracuse Research Corporation, Ft. Washington, PA
FRANZ OESCH, University of Mainz, Mainz, Germany
RICHARD B. SCHLESINGER, Pace University, Pleasantville, NY
CALVIN C. WILLHITE, Department of Toxic Substances, State of California, Berkeley
FREDERIK A. DE WOLFF, Leiden University, Leiden, The Netherlands

Staff

KULBIR S. BAKSHI, Program Director
KELLY CLARK, Editor
AIDA C. NEEL, Senior Project Assistant

COMMITTEE ON TOXICOLOGY

Members

BAILUS WALKER, JR. *(Chair)*, Howard University Medical Center and American Public Health Association, Washington, DC
MELVIN E. ANDERSEN, CIIT-Centers for Health Research, Research Triangle Park, NC
EDWARD C. BISHOP, Parsons Corporation, Pasadena, CA
GARY P. CARLSON, Purdue University, West Lafayette, IN
JANICE E. CHAMBERS, Mississippi State University, Mississippi State
LEONARD CHIAZZE, JR., Georgetown University, Washington, DC
JUDITH A. GRAHAM, American Chemistry Council, Arlington, VA
SIDNEY GREEN, Howard University, Washington, DC
MERYL KAROL, University of Pittsburgh, Pittsburgh, PA
STEPHEN U. LESTER, Center for Health Environment and Justice, Falls Church, VA
DAVID H. MOORE, Battelle Memorial Institute, Bel Air, MD
CALVIN C. WILLHITE, Department of Toxic Substances, State of California, Berkeley
GERALD WOGAN, Massachusetts Institute of Technology, Cambridge

Staff

KULBIR S. BAKSHI, Program Director for Toxicology
ROBERTA M. WEDGE, Program Director for Risk Analysis
SUSAN N.J. MARTEL, Senior Staff Officer
ELLEN K. MANTUS, Senior Staff Officer
KELLY CLARK, Assistant Editor
AIDA C. NEEL, Senior Project Assistant
TAMARA DAWSON, Project Assistant

---

[1]This study was planned, overseen, and supported by the Board on Environmental Studies and Toxicology.

OTHER REPORTS OF THE BOARD ON ENVIRONMENTAL STUDIES AND TOXICOLOGY

Air Quality Management in the United States (2004)
Endangered and Threatened Fishes in the Klamath River Basin: Causes of Decline and Strategies for Recovery (2004)
Cumulative Environmental Effects of Alaska North Slope Oil and Gas Development (2003)
Estimating the Public Health Benefits of Proposed Air Pollution Regulations (2002)
Biosolids Applied to Land: Advancing Standards and Practices (2002)
The Airliner Cabin Environment and Health of Passengers and Crew (2002)
Arsenic in Drinking Water: 2001 Update (2001)
Evaluating Vehicle Emissions Inspection and Maintenance Programs (2001)
Compensating for Wetland Losses Under the Clean Water Act (2001)
A Risk-Management Strategy for PCB-Contaminated Sediments (2001)
Acute Exposure Guideline Levels for Selected Airborne Chemicals (4 volumes, 2000-2004)
Toxicological Effects of Methylmercury (2000)
Strengthening Science at the U.S. Environmental Protection Agency (2000)
Scientific Frontiers in Developmental Toxicology and Risk Assessment (2000)
Ecological Indicators for the Nation (2000)
Waste Incineration and Public Health (1999)
Hormonally Active Agents in the Environment (1999)
Research Priorities for Airborne Particulate Matter (4 volumes, 1998-2003)
Arsenic in Drinking Water (1999)
The National Research Council's Committee on Toxicology: The First 50 Years (1997)
Carcinogens and Anticarcinogens in the Human Diet (1996)
Upstream: Salmon and Society in the Pacific Northwest (1996)
Science and the Endangered Species Act (1995)
Wetlands: Characteristics and Boundaries (1995)
Biologic Markers (5 volumes, 1989-1995)
Review of EPA's Environmental Monitoring and Assessment Program (3 volumes, 1994-1995)
Science and Judgment in Risk Assessment (1994)
Pesticides in the Diets of Infants and Children (1993)
Dolphins and the Tuna Industry (1992)
Science and the National Parks (1992)
Human Exposure Assessment for Airborne Pollutants (1991)
Rethinking the Ozone Problem in Urban and Regional Air Pollution (1991)
Decline of the Sea Turtles (1990)

Copies of these reports may be ordered from the National Academies Press
(800) 624-6242 or (202) 334-3313
www.nap.edu

OTHER REPORTS OF THE COMMITTEE ON TOXICOLOGY

Spacecraft Water Exposure Guidelines for Selected Contaminants, Volume 1 (2004)
Toxicologic Assessment of Jet-Propulsion Fuel 8 (2003)
Review of Submarine Escape Action Levels for Selected Chemicals (2002)
Standing Operating Procedures for Developing Acute Exposure Guideline Levels for Hazardous Chemicals (2001)
Evaluating Chemical and Other Agent Exposures for Reproductive and Developmental Toxicity (2001)
Acute Exposure Guideline Levels for Selected Airborne Contaminants, Volume 1 (2000), Volume 2 (2002), Volume 3 (2003), Volume 4 (2004)
Review of the US Navy's Human Health Risk Assessment of the Naval Air Facility at Atsugi, Japan (2000)
Methods for Developing Spacecraft Water Exposure Guidelines (2000)
Review of the U.S. Navy Environmental Health Center's Health-Hazard Assessment Process (2000)
Review of the U.S. Navy's Exposure Standard for Manufactured Vitreous Fibers (2000)
Re-Evaluation of Drinking-Water Guidelines for Diisopropyl Methylphosphonate (2000)
Submarine Exposure Guidance Levels for Selected Hydrofluorocarbons: HFC-236fa, HFC-23, and HFC-404a (2000)
Review of the U.S. Army's Health Risk Assessments for Oral Exposure to Six Chemical-Warfare Agents (1999)
Toxicity of Military Smokes and Obscurants, Volume 1(1997), Volume 2 (1999), Volume 3 (1999)
Assessment of Exposure-Response Functions for Rocket-Emission Toxicants (1998)
Toxicity of Alternatives to Chlorofluorocarbons: HFC-134a and HCFC-123 (1996)
Permissible Exposure Levels for Selected Military Fuel Vapors (1996)
Spacecraft Maximum Allowable Concentrations for Selected Airborne Contaminants, Volume 1 (1994), Volume 2 (1996), Volume 3 (1996), Volume 4 (2000)

Preface

Extremely hazardous substances (EHSs)[1] can be released accidentally as a result of chemical spills, industrial explosions, fires, or accidents involving railroad cars and trucks transporting EHSs. The people in communities surrounding industrial facilities where EHSs are manufactured, used, or stored and in communities along the nation's railways and highways potentially are at risk of being exposed to airborne EHSs during accidental releases. Pursuant to the Superfund Amendments and Reauthorization Act of 1986, the U.S. Environmental Protection Agency (EPA) has identified approximately 400 EHSs on the basis of acute lethality data in rodents.

As part of its efforts to develop acute exposure guideline levels for EHSs, EPA and the Agency for Toxic Substances and Disease Registry (ATSDR) in 1991 requested that the National Research Council (NRC) develop guidelines for establishing such levels. In response to that request, the NRC published *Guidelines for Developing Community Emergency Exposure Levels for Hazardous Substances* in 1993.

Using the 1993 NRC guidelines report, the National Advisory Committee (NAC) on Acute Exposure Guideline Levels for Hazardous Substances —consisting of members from EPA, the Department of Defense (DOD), the Department of Energy (DOE), the Department of Transportation, other federal and state governments, the chemical industry, academe, and other

[1]As defined pursuant to the Superfund Amendments and Reauthorization Act of 1986.

organizations from the private sector—has developed acute exposure guideline levels (AEGLs) for approximately 80 EHSs.

In 1998, EPA and DOD requested that the NRC independently review the AEGLs developed by NAC. In response to that request, the NRC organized within its Committee on Toxicology the Subcommittee on Acute Exposure Guideline Levels, which prepared this report. This report is the fourth volume in the series *Acute Exposure Guideline Levels for Selected Airborne Chemicals*. It reviews the AEGLs for chlorine, hydrogen chloride, hydrogen fluoride, toluene 2,4- and 2,6-diisocyanate, and uranium hexafluoride for scientific accuracy, completeness, and consistency with the NRC guideline reports.

This report was reviewed in draft by individuals selected for their diverse perspectives and technical expertise, in accordance with procedures approved by the NRC's Report Review Committee. The purpose of this independent review is to provide candid and critical comments that will assist the institution in making its published report as sound as possible and to ensure that the report meets institutional standards for objectivity, evidence, and responsiveness to the study charge. The review comments and draft manuscript remain confidential to protect the integrity of the deliberative process. We wish to thank the following individuals for their review of this report: David H. Moore of Battelle Memorial Institute; Sam Kacew of University of Ottawa; and Rakesh Dixit of Merck and Company, Inc.

Although the reviewers listed above have provided many constructive comments and suggestions, they were not asked to endorse the conclusions or recommendations nor did they see the final draft of the report before its release. The review of this report was overseen by Janice E. Chambers of Mississippi State University, appointed by the Division on Earth and Life Studies, who was responsible for making certain that an independent examination of this report was carried out in accordance with institutional procedures and that all review comments were carefully considered. Responsibility for the final content of this report rests entirely with the authoring committee and the institution.

The subcommittee gratefully acknowledges the valuable assistance provided by the following people: Ernest Falke and Paul Tobin, EPA; George Rusch, Honeywell, Inc.; Sylvia Talmage, Cheryl Bast, and Carol Wood, Oak Ridge National Laboratory; and Aida Neel, senior project assistant for the Board on Environmental Studies and Toxicology. Kelly Clark edited the report. We are grateful to James J. Reisa, director of the Board on Environmental Studies and Toxicology, for his helpful comments. The subcommittee particularly acknowledges Kulbir Bakshi, project director for the subcommittee, for bringing the report to completion. Finally, we would like to thank all members of the subcommittee for their expertise and dedicated effort throughout the development of this report.

Daniel Krewski, *Chair*
Subcommittee on Acute Exposure Guideline Levels

Bailus Walker, *Chair*
Committee on Toxicology

Contents

Acute Exposure Guideline Levels for Selected Airborne Chemicals

Volume 4

Introduction

This report is the fourth volume in the series *Acute Exposure Guideline Levels for Selected Airborne Chemicals.*

In the Bhopal disaster of 1984, approximately 2,000 residents living near a chemical plant were killed and 20,000 more suffered irreversible damage to their eyes and lungs following accidental release of methyl isocyanate. The toll was particularly high because the community had little idea what chemicals were being used at the plant, how dangerous they might be, and what steps to take in case of emergency. This tragedy served to focus international attention on the need for governments to identify hazardous substances and to assist local communities in planning how to deal with emergency exposures.

In the United States, the Superfund Amendments and Reauthorization Act (SARA) of 1986 required that the U.S. Environmental Protection Agency (EPA) identify extremely hazardous substances (EHSs) and, in cooperation with the Federal Emergency Management Agency and the Department of Transportation, assist Local Emergency Planning Committees (LEPCs) by providing guidance for conducting health-hazard assessments for the development of emergency-response plans for sites where EHSs are produced, stored, transported, or used. SARA also required that the Agency for Toxic Substances and Disease Registry (ATSDR) determine whether chemical substances identified at hazardous waste sites or in the environment present a public-health concern.

As a first step in assisting the LEPCs, EPA identified approximately 400 EHSs largely on the basis of their immediately dangerous to life and health (IDLH) values developed by the National Institute for Occupational Safety and Health (NIOSH) in experimental animals. Although several public and private groups, such as the Occupational Safety and Health Administration (OSHA) and the American Conference of Governmental Industrial Hygienists (ACGIH), have established exposure limits for some substances and some exposures (e.g., workplace or ambient air quality), these limits are not easily or directly translated into emergency exposure limits for exposures at high levels but of short duration, usually less than 1 h, and only once in a lifetime for the general population, which includes infants (from birth to 3 years of age), children, the elderly, and persons with diseases, such as asthma or heart disease.

The National Research Council (NRC) Committee on Toxicology (COT) has published many reports on emergency exposure guidance levels and spacecraft maximum allowable concentrations for chemicals used by the Department of Defense (DOD) and the National Aeronautics and Space Administration (NASA) (NRC 1968, 1972, 1984a,b,c,d, 1985a,b, 1986a,b, 1987, 1988, 1994, 1996a,b, 2000). COT has also published guidelines for developing emergency exposure guidance levels for military personnel and for astronauts (NRC 1986b, 1992). Because of COT's experience in recommending emergency exposure levels for short-term exposures, in 1991 EPA and ATSDR requested that COT develop criteria and methods for developing emergency exposure levels for EHSs for the general population. In response to that request, the NRC assigned this project to the COT Subcommittee on Guidelines for Developing Community Emergency Exposure Levels for Hazardous Substances. The report of that subcommittee, *Guidelines for Developing Community Emergency Exposure Levels for Hazardous Substances* (NRC 1993), provides step-by-step guidance for setting emergency exposure levels for EHSs. Guidance is given on what data are needed, what data are available, how to evaluate the data, and how to present the results.

In November1995, the National Advisory Committee for Acute Exposure Guideline Levels for Hazardous Substances (NAC)[1] was established to identify, review, and interpret relevant toxicologic and other scientific data and to develop acute exposure guideline levels (AEGLs) for high-priority, acutely toxic chemicals. The NRC's previous name for acute exposure levels—community emergency exposure levels (CEELs)—was re-

[1]NAC is composed of members from EPA, DOD, many other federal and state agencies, industry, academe, and other organizations. The roster of NAC is shown on page 8.

placed by "AEGLs" to reflect the broad application of these values to planning, response, and prevention in the community, the workplace, transportation, the military, and the remediation of Superfund sites.

AEGLs represent threshold exposure limits (exposure levels below which adverse health effects are not likely to occur) for the general public and are applicable to emergency exposures ranging from 10 min to 8 h. Three levels—AEGL-1, AEGL-2, and AEGL-3—are developed for each of five exposure periods (10 min, 30 min, 1 h, 4 h, and 8 h) and are distinguished by varying degrees of severity of toxic effects. The three AEGLs are defined as follows:

> AEGL-1 is the airborne concentration (expressed as ppm [parts per million] or mg/m^3 [milligrams per cubic meter]) of a substance above which it is predicted that the general population, including susceptible individuals, could experience notable discomfort, irritation, or certain asymptomatic nonsensory effects. However, the effects are not disabling and are transient and reversible upon cessation of exposure.
>
> AEGL-2 is the airborne concentration (expressed as ppm or mg/m^3) of a substance above which it is predicted that the general population, including susceptible individuals, could experience irreversible or other serious, long-lasting adverse health effects or an impaired ability to escape.
>
> AEGL-3 is the airborne concentration (expressed as ppm or mg/m^3) of a substance above which it is predicted that the general population, including susceptible individuals, could experience life-threatening adverse health effects or death.

Airborne concentrations below AEGL-1 represent exposure levels that can produce mild and progressively increasing but transient and nondisabling odor, taste, and sensory irritation or certain asymptomatic, nonsensory adverse effects. With increasing airborne concentrations above each AEGL, there is a progressive increase in the likelihood of occurrence and the severity of effects described for each corresponding AEGL. Although the AEGL values represent threshold levels for the general public, including susceptible subpopulations, such as infants, children, the elderly, persons with asthma, and those with other illnesses, it is recognized that individuals, subject to unique or idiosyncratic responses, could experience the effects described at concentrations below the corresponding AEGL.

SUMMARY OF REPORT ON GUIDELINES FOR DEVELOPING AEGLS

As described in the *Guidelines for Developing Community Emergency Exposure Levels for Hazardous Substances* (NRC 1993) and the NAC guidelines report *Standing Operating Procedures on Acute Exposure Guideline Levels for Hazardous Substances*(NRC 2001), the first step in establishing AEGLs for a chemical is to collect and review all relevant published and unpublished information available on that chemical. Various types of evidence are assessed in establishing AEGL values for a chemical. They include information from (1) chemical-physical characterizations, (2) structure-activity relationships, (3) in vitro toxicity studies, (4) animal toxicity studies, (5) controlled human studies, (6) observations of humans involved in chemical accidents, and (7) epidemiologic studies. Toxicity data from human studies are most applicable and are used when available in preference to data from animal studies and in vitro studies. Toxicity data from inhalation exposures are most useful for setting AEGLs for airborne chemicals because inhalation is the most likely route of exposure and because extrapolation of data from other routes would lead to additional uncertainty in the AEGL estimate.

For most chemicals, actual human toxicity data are not available or critical information on exposure is lacking, so toxicity data from studies conducted in laboratory animals are extrapolated to estimate the potential toxicity in humans. Such extrapolation requires experienced scientific judgment. The toxicity data from animal species most representative of humans in terms of pharmacodynamic and pharmacokinetic properties are used for determining AEGLs. If data are not available on the species that best represents humans, the data from the most sensitive animal species are used to set AEGLs. Uncertainty factors are commonly used when animal data are used to estimate risk levels for humans. The magnitude of uncertainty factors depends on the quality of the animal data used to determine the no-observed-adverse-effect level (NOAEL) and the mode of action of the substance in question. When available, pharmacokinetic data on tissue doses are considered for interspecies extrapolation.

For substances that affect several organ systems or have multiple effects, all end points, including reproductive (in both genders), developmental, neurotoxic, respiratory, and other organ-related effects, are evaluated, the most important or most sensitive effect receiving the greatest attention. For carcinogenic chemicals, excess carcinogenic risk is estimated, and the AEGLs corresponding to carcinogenic risks of 1 in 10,000 (1×10^{-4}), 1 in

100,000 (1×10^{-5}), and 1 in 1,000,000 (1×10^{-6}) exposed persons are estimated.

REVIEW OF AEGL REPORTS

As NAC began developing chemical-specific AEGL reports, EPA and DOD asked the NRC to review independently the NAC reports for their scientific validity, completeness, and consistency with the NRC guideline reports (NRC 1993, 2001). The NRC assigned this project to the COT Subcommittee on Acute Exposure Guideline Levels. The subcommittee has expertise in toxicology, epidemiology, pharmacology, medicine, industrial hygiene, biostatistics, risk assessment, and risk communication.

The AEGL draft reports are initially prepared by ad hoc AEGL Development Teams consisting of a chemical manager, two chemical reviewers, and a staff scientist of the NAC contractor—Oak Ridge National Laboratory. The draft documents are then reviewed by NAC and elevated from "draft" to "proposed" status. After the AEGL documents are approved by NAC, they are published in the *Federal Register* for public comment. The reports are then revised by NAC in response to the public comments, elevated from "proposed" to "interim" status, and sent to the NRC Subcommittee on Acute Exposure Guideline Levels for final evaluation.

The NRC subcommittee's review of the AEGL reports prepared by NAC and its contractors involves oral and written presentations to the subcommittee by the authors of the reports. The NRC subcommittee provides advice and recommendations for revisions to ensure scientific validity and consistency with the NRC guideline reports (NRC 1993, 2001). The revised reports are presented at subsequent meetings until the subcommittee is satisfied with the reviews.

Because of the enormous amount of data presented in the AEGL reports, the NRC subcommittee cannot verify all the data used by NAC. The NRC subcommittee relies on NAC for the accuracy and completeness of the toxicity data cited in the AEGLs reports.

This report is the fourth volume in the series *Acute Exposure Guideline Levels for Selected Airborne Chemicals*. AEGL documents for chlorine, hydrogen chloride, hydrogen fluoride, toluene 2,4- and 2,6-diisocyanate, and uranium hexafluoride are published as an appendix to this report. The subcommittee concludes that the AEGLs developed in those documents are scientifically valid conclusions based on the data reviewed by NAC and are consistent with the NRC guideline reports. AEGL reports for additional chemicals will be presented in subsequent volumes.

REFERENCES

NRC (National Research Council). 1968. Atmospheric Contaminants in Spacecraft. Washington, DC: National Academy of Sciences.

NRC (National Research Council). 1972. Atmospheric Contaminants in Manned Spacecraft. Washington, DC: National Academy of Sciences.

NRC (National Research Council). 1984a. Emergency and Continuous Exposure Limits for Selected Airborne Contaminants, Vol. 1. Washington, DC: National Academy Press.

NRC (National Research Council). 1984b. Emergency and Continuous Exposure Limits for Selected Airborne Contaminants, Vol. 2. Washington, DC: National Academy Press.

NRC (National Research Council). 1984c. Emergency and Continuous Exposure Limits for Selected Airborne Contaminants, Vol. 3. Washington, DC: National Academy Press.

NRC (National Research Council). 1984d. Toxicity Testing: Strategies to Determine Needs and Priorities. Washington, DC: National Academy Press.

NRC (National Research Council). 1985a. Emergency and Continuous Exposure Guidance Levels for Selected Airborne Contaminants, Vol. 4. Washington, DC: National Academy Press.

NRC (National Research Council). 1985b. Emergency and Continuous Exposure Guidance Levels for Selected Airborne Contaminants, Vol. 5. Washington, DC: National Academy Press.

NRC (National Research Council). 1986a. Emergency and Continuous Exposure Guidance Levels for Selected Airborne Contaminants, Vol. 6. Washington, DC: National Academy Press.

NRC (National Research Council). 1986b. Criteria and Methods for Preparing Emergency Exposure Guidance Level (EEGL), Short-Term Public Emergency Guidance Level (SPEGL), and Continuous Exposure Guidance level (CEGL) Documents. Washington, DC: National Academy Press.

NRC (National Research Council). 1987. Emergency and Continuous Exposure Guidance Levels for Selected Airborne Contaminants, Vol. 7. Washington, DC: National Academy Press.

NRC (National Research Council). 1988. Emergency and Continuous Exposure Guidance Levels for Selected Airborne Contaminants, Vol. 8. Washington, DC: National Academy Press.

NRC (National Research Council). 1992. Guidelines for Developing Spacecraft Maximum Allowable Concentrations for Space Station Contaminants. Washington, DC: National Academy Press.

NRC (National Research Council). 1993. Guidelines for Developing Community Emergency Exposure Levels for Hazardous Substances. Washington, DC: National Academy Press.

NRC (National Research Council). 1994. Spacecraft Maximum Allowable Concentrations for Selected Airborne Contaminants, Vol. 1. Washington, DC: National Academy Press.

NRC (National Research Council). 1996a. Spacecraft Maximum Allowable Concentrations for Selected Airborne Contaminants, Vol. 2. Washington, DC: National Academy Press.

NRC (National Research Council). 1996b. Spacecraft Maximum Allowable Concentrations for Selected Airborne Contaminants, Vol. 3. Washington, DC: National Academy Press.

NRC (National Research Council). 2000. Spacecraft Maximum Allowable Concentrations for Selected Airborne Contaminants, Vol. 4. Washington, DC: National Academy Press.

NRC (National Research Council) 2001. Acute Exposure Guideline Levels for Selected Airborne Chemicals. Volume 1. Washington, DC: National Academy Press.

NRC (National Research Council). 2001. Standing Operating Procedures for Developing Acute Exposure Guideline Levels for Airborne Chemicals. Washington, DC: National Academy Press.

NRC (National Research Council) 2002. Acute Exposure Guideline Levels for Selected Airborne Chemicals. Volume 2. Washington, DC: National Academy Press.

NRC (National Research Council) 2003. Acute Exposure Guideline Levels for Selected Airborne Chemicals. Volume 3. Washington, DC: National Academy Press.

Roster of the National Advisory Committee for Acute Exposure Guideline Levels for Hazardous Substances

Committee Members

George Rusch
Chair, NAC/AEGL Committee
Department of Toxicology and Risk Assessment
Honeywell, Inc.
Morristown, NJ

Steven Barbee
Arch Chemicals, Inc.
Norwalk, CT

Ernest Falke
Chair, SOP Workgroup
U.S. Environmental Protection Agency
Washington, DC

Lynn Beasley
U.S. Environmental Protection Agency
Washington, DC

George Alexeeff
Office of Environmental Health Hazard Assessment
California EPA
Oakland, CA

Robert Benson
U.S. Environmental Protection Agency
Region VIII
Denver, CO

Jonathan Borak
Yale University
New Haven, CT

William Bress
Vermont Department of Health
Burlington, VT

George Cushmac
Office of Hazardous Materials Safety
U.S. Department of Transporation
Washington, DC

Larry Gephart
Exxon Mobil Biomedical Sciences
Annandale, NJ

John P. Hinz
U.S. Air Force
Brooks Air Force Base, TX

James Holler
Agency for Toxic Substances and Disease Registry
Atlanta, GA

Thomas C. Hornshaw
Office of Chemical Safety
Illinois Environmental Protection Agency
Springfield, IL

Nancy K. Kim
Division of Environmental Health Assessment
New York State Department of Health
Troy, NY

Loren Koller
Loren Koller & Associates
Corvallis, OR

Glenn Leach
U.S. Army Center for Health Promotion and Preventive Medicine
Aberdeen Proving Grounds, MD

John Morawetz
International Chemical Workers Union
Cincinnati, OH

Richard W. Niemeier
National Institute for Occupational Safety and Health
Cincinnati, OH

Marinelle Payton
Department of Public Health
Jackson State University
Jackson, MS

George Rodgers
Department of Pediatrics
Division of Critical Care
University of Louisville
Louisville, KY

Robert Snyder
Environmental and Occupational Health Sciences Institute
Piscataway, NJ

Thomas J. Sobotka
U.S. Food and Drug Administration
Laurel, MD

Richard Thomas
International Center for Environmental Technology
McLean, VA

Oak Ridge National Laboratory Staff

Cheryl Bast
Oak Ridge National Laboratory
Oak Ridge, TN

Carol Wood
Oak Ridge National Laboratory
Oak Ridge, TN

Sylvia Talmage
Oak Ridge National Laboratory
Oak Ridge, TN

National Advisory Committee Staff

Paul S. Tobin
Designated Federal Officer, AEGL Program
U.S. Environmental Protection Agency
Washington, DC

Marquea King
Senior Scientist
U.S. Environmental Protection Agency
Washington, DC

Appendixes

1

Chlorine[1]

Acute Exposure Guideline Levels

SUMMARY

Chlorine is a greenish-yellow, highly reactive halogen gas that has a pungent, suffocating odor. The vapor is heavier than air and will form a cloud in the vicinity of a spill. Like other halogens, chlorine exists in the diatomic state in nature. Chlorine is extremely reactive and rapidly combines with both inorganic and organic substances. Chlorine is used in the manufacture of a wide variety of chemicals, as a bleaching agent in industry and household products, and as a biocide in water and waste treatment plants.

[1]This document was prepared by the AEGL Development Team comprising Sylvia Talmage (Oak Ridge National Laboratory) and members of the National Advisory Committee (NAC) on Acute Exposure Guideline Levels for Hazardous Substances, including Larry Gephart (Chemical Manager) and George Alexeeff and Kyle Blackman (Chemical Reviewers). The NAC reviewed and revised the document and AEGLs as deemed necessary. Both the document and the AEGL values were then reviewed by the National Research Council (NRC) Subcommittee on Acute Exposure Guideline Levels. The NRC subcommittee concludes that the AEGLs developed in this document are scientifically valid on the basis of the data reviewed by the NRC and are consistent with the NRC guidelines reports (NRC 1993, 2001).

Chlorine is an irritant to the eyes and respiratory tract; reaction with moist surfaces produces hydrochloric and hypochlorous acids. Its irritant properties have been studied in human volunteers, and its acute inhalation toxicity has been studied in several laboratory animal species. The data from the human and laboratory animal studies were sufficient for developing acute exposure guideline levels (AEGLs) for the five exposure durations (i.e., 10 and 30 minutes [min] and 1, 4, and 8 hours [h]). Regression analysis of human data on nuisance irritation responses (itching or burning of the eyes, nose, or throat) for durations of 30-120 min and during exposures to chlorine at 0-2 parts per million (ppm) determined that the relationship between concentration and time is approximately $C^2 \times t = k$ (where C = concentration, t = time, and k is a constant) (ten Berge and Vis van Heemst 1983).

The AEGL-1 was based on a combination of studies that tested healthy human subjects as well as atopic individuals (Rotman et al. 1983; Shusterman et al. 1998) and asthmatic patients (D'Alessandro et al. 1996). Atopic and asthmatic individuals have been identified as susceptible populations for irritant gases. The highest no-observed-adverse-effect level (NOAEL) for notable irritation and significant changes in pulmonary function parameters was 0.5 ppm in two studies. Eight atopic subjects were exposed for 15 min in one study (Shusterman et al. 1998), and eight healthy exercising individuals and an exercising atopic individual were exposed for two consecutive 4-h periods in the other (Rotman et al. 1983). The subjects in the Shusterman et al. (1998) study experienced nasal congestion, but irritation was described as none to slight. The exercising atopic individual in the Rotman et al. (1983) study experienced nondisabling, transient, asymptomatic changes in pulmonary function parameters. The selection of 0.5 ppm is supported by the lack of symptoms and lack of changes in pulmonary air flow and airway resistance in five asthmatic subjects inhaling 0.4 ppm for 1 h (D'Alessandro et al. 1996).

Because susceptible populations comprising atopic and asthmatic individuals were tested at similar concentrations, with incorporation of exercise into the protocol of one study, an intraspecies uncertainty factor (UF) of 1 was applied. The intraspecies UF of 1 is further supported by the fact that pediatric asthmatic subjects do not appear to be more responsive to irritants than adult asthmatic subjects (Avital et al. 1991). The AEGL-1 value was not time scaled for several reasons. First, the Rotman et al. (1983) study was for 8 h with a single 1-h break. Second, the response to chlorine appears to be concentration-dependent rather than time-dependent, as the pulmonary function parameters of individuals tested in this study, including those for the atopic individual, did not increase between the 4- and 8-h measurements.

The AEGL-2 values were based on two of the studies used to derive the AEGL-1. Both healthy and susceptible human subjects inhaled chlorine at 1.0 ppm for 1 h (D'Alessandro et al. 1996) or 4 h (Rotman et al. 1983). Both healthy and susceptible subjects experienced some sensory irritation and transient changes in pulmonary function measurements. Greater changes were observed in pulmonary parameters among the susceptible subjects compared with the normal groups. In the latter study (Rotman et al. 1983), an atopic individual experienced no respiratory symptoms other than some sensory irritation during the 4-h exposure, but his airway resistance nearly tripled. He experienced shortness of breath and wheezing during a second 4-h exposure. Five individuals with nonspecific airway hyper-reactivity or asthma also experienced a statistically significant fall in pulmonary air flow and an increase in airway resistance during a 1-h exposure at 1.0 ppm (D'Alessandro et al. (1996). There were no respiratory symptoms during the exposure. The susceptible individual in the Rotman et al. (1983) study remained in the exposure chamber for the full 4 h without respiratory symptoms. Therefore, when considering the definition of the AEGL-2, the first 4 h of exposure was a no-effect level in a susceptible individual. Because the subjects were susceptible individuals, one of the subjects was undergoing light exercise during the exposures (making him more vulnerable to sensory effects), and an exercising susceptible individual exhibited effects that did not impede escape for the 4-h exposure duration (consistent with the definition of the AEGL-2), an intraspecies UF of 1 was applied.

Chlorine is a highly irritating and corrosive gas that reacts directly with the tissues of the respiratory tract with no pharmacokinetic component involved in toxicity; therefore, effects are not expected to vary greatly among other susceptible populations. Time-scaling was considered appropriate for the AEGL-2 because it is defined as the threshold for irreversible effects, which, in the case of irritants, generally involves tissue damage. Although the end point used in this case—a no-effect concentration for wheezing that was accompanied by a significant increase in airways resistance—has a different mechanism of action than that of direct tissue damage, it is assumed that some biomarkers of tissue irritation would be present in the airways and lungs. The 4-h 1-ppm concentration was scaled to the other time periods using the $C^2 \times t = k$ relationship. The scaling factor was based on regression analyses of concentrations and exposure durations that attained nuisance levels of irritation in human subjects (ten Berge and Vis van Heemst 1983). The 10-min value was set equal to the 30-min value in order to not exceed the highest exposure of 4.0 ppm in controlled human studies.

In the absence of human data, animal lethality data served as the basis for AEGL-3. The mouse was not chosen as an appropriate model for lethality because mice often showed delayed deaths, which several authors attributed to bronchopneumonia. Because the mouse was shown to be more sensitive to chlorine than the dog and rat, and because the mouse does not provide an appropriate basis for quantitatively predicting mortality in humans, a value below those resulting in no deaths in the rat (213 ppm and 322 ppm) and above that resulting in no deaths in the mouse (150 ppm) for a period of 1 h was chosen (MacEwen and Vernot 1972; Zwart and Woutersen 1988). The AEGL-3 values were derived from a 1-h concentration of 200 ppm. That value was calculated applying a total UF of 10—3 to extrapolate from rats to humans (interspecies values for the same end point differed by a factor of approximately 2 within each of several studies), and 3 to account for differences in human sensitivity. The susceptibility of asthmatic subjects relative to healthy subjects when considering lethality is unknown, but the data from two studies with human subjects showed that doubling a no-effect concentration for irritation and bronchial constriction resulted in potentially serious effects in asthmatic subjects but not in normal individuals. Time-scaling was considered appropriate for the AEGL-3, because tissue damage is involved. (Data in animal studies clearly indicate that time scaling is appropriate when lung damage is observed.) The AEGL-3 values for the other exposure times were calculated using the $C^2 \times t = k$ relationship, which was derived based on the end point of irritation from a study with humans.

The calculated values are listed in Table 1-1.

1. INTRODUCTION

Chlorine is the most abundant naturally occurring halogen. Halogens do not occur in the elemental state in nature. When formed experimentally, chlorine is a greenish-yellow, diatomic gas (Cl_2) with a pungent, suffocating odor. Chlorine is used in the manufacture of a variety of nonagricultural chemicals, such as vinyl chloride and ethylene dichloride; as a bleaching agent in the paper industry (along with chlorine dioxide [ClO_2]); as commercial and household bleaching agents (in the form of chlorates [ClO_3^-] and hypochlorites [OCl^-]); and as a biocide in water purification and waste

TABLE 1-1 Summary of AEGLs Values for Chlorine (ppm [mg/m^3])

Classification	10 min	30 min	1 h	4 h	8 h	End Point (Reference)
AEGL-1[a] (Nondisabling)	0.5 (1.5)	0.5 (1.5)	0.5 (1.5)	0.5 (1.5)	0.5[b] (1.5)	No to slight changes in pulmonary function parameters in humans (Rotman et al. 1983; D'Alessandro et al. 1996; Shusterman et al. 1998)
AEGL-2 (Disabling)	2.8 (8.1)	2.8 (8.1)	2.0 (5.8)	1.0 (2.9)	0.7 (2.0)	1.0 ppm for 4 h was a NOAEL for an asthma-like attack in human subjects; the other values were time-scaled (Rotman et al. 1983; D'Alessandro et al. 1996)
AEGL-3 (Lethal)	50 (145)	28 (81)	20 (58)	10 (29)	7.1 (21)	Threshold for lethality in the rat (MacEwen and Vernot 1972; Zwart and Woutersen 1988)

[a]The distinctive, pungent odor of chlorine will be noticeable to most individuals at these concentrations.
[b]Because effects were not increased following an interrupted 8-h exposure of an atopic individual to 0.5 ppm, the 8-h AEGL-1 was set equal to 0.5 ppm.
Abbreviations: mg/m^3, milligrams per cubic meter; ppm, parts per million.

treatment systems (Perry et al. 1994). Chlorine gas was used as a chemical warfare agent during World War I (Withers and Lees 1987). The vapor is heavier than air and will form a cloud in the vicinity of a spill.

As of January 1999, world annual capacity for chlorine production was estimated at almost 50 million metric tons (CEH 2000). Chlorine is produced at chlor-alkali plants at over 650 sites worldwide, and North America accounts for 32% of capacity (operating rates are greater than 83% of capacity). In the early 1990s, chlorine was produced at 49 facilities, operated by 29 companies, in the United States (Perry et al. 1994). In 1993, U.S.

production was reported at 24 billion pounds (C&EN 1994). The major global market for chlorine is ethylene dichloride production (about 33%) (CEH 2000).

Chlorine is extremely reactive and enters into substitution or addition reactions with both inorganic and organic substances. Moist chlorine unites directly with most elements. Reaction with water produces hydrochloric (HCl) and hypochlorous acid (HClO) (Budavari et al. 1996; Perry et al. 1994). Other relevant chemical and physical properties are listed in Table 1-2. According to Amoore and Hautala (1983), the odor threshold is 0.31 ppm, and a range of 0.2-0.4 ppm was reported in other studies. There is considerable variation in detecting the odor among subjects; for many individuals, the ability to perceive the odor decreases over exposure time (NIOSH 1976).

Chlorine is an eye and respiratory tract irritant and, at high doses, has direct toxic effects on the lungs. It reaches the lungs because it is only moderately soluble in water and it is not totally absorbed in the upper respiratory tract at high concentrations. The acute inhalation toxicity of chlorine has been studied in several laboratory animal species, and its irritant properties have been studied with human volunteers.

2. HUMAN TOXICITY DATA

2.1 Acute Lethality

For humans, a 5-min lethal concentration in 10% of subjects (LC_{10}) of 500 ppm (NTIS 1996) and a possible 30-min lethal exposure of 872 ppm have been reported (Perry et al. 1995). Both of those secondary sources cited data from Prentiss (1937) as well as data from other early sources.

Although accidental releases have resulted in deaths (e.g., Jones et al. 1986), no studies were located in which acute lethal exposure concentrations were measured. Probit analysis of available information on the lethality of chlorine to animals and humans was used by Withers and Lees (1985b) to estimate a concentration lethal to 50% of the population (LC_{50}). Their model incorporates the effects of physical activity, inhalation rate, the effectiveness of medical treatment, and the lethal toxic load function. The estimated 30-min LC_{50} at a standard level of activity (inhalation rate of 12 liters [L]/min) for the regular, vulnerable, and average (regular plus vulnerable) populations, as described by the authors, were 250, 100, and 210 ppm,

TABLE 1-2 Chemical and Physical Properties of Chlorine

Parameter	Value	Reference
Synonyms	Bertholite; hypochlorite; hypochlorous acid	Budavari et al. 1996
Molecular formula	Cl_2	Budavari et al. 1996
Molecular weight	70.9	Budavari et al. 1996
CAS registry no.	7782-50-5	Budavari et al. 1996
Physical state	Gas	Budavari et al. 1996
Color	Greenish-yellow	Budavari et al. 1996
Solubility in water	0.092 moles/L	Budavari et al. 1996
Vapor pressure	5,025 mm Hg at 20°C	Matheson Gas Co. 1980
Vapor density	1.4085 at 20°C	AIHA 1988
Density (water =1)	1.56 at boiling point	Perry et al. 1994
Melting point	-101°C	Budavari et al. 1996
Boiling point	-34.05°C	Budavari et al. 1996
Flammability	Nonflammable	Matheson Gas Co. 1980
Conversion factors in air	1 ppm = 2.9 mg/m^3 1 mg/m^3 = 0.34 ppm	ACGIH 2001

respectively. The LC_{10} for the three populations were 125, 50, and 80 ppm, respectively.

2.2. Nonlethal Toxicity

Exposures at 30 ppm and 40-60 ppm have been reported to cause intense coughing and serious damage, respectively (ILO 1998), but no documentation of those values was given.

2.2.1. Experimental Studies

Five well-conducted and well-documented studies using human volunteers were located. Those studies are summarized in Table 1-3.

TABLE 1-3 Summary of Irritant Effects in Humans[a]

Concentration (ppm)	Exposure Time[b]	Effect	References
0.4	1 h	No pulmonary function changes in subjects with airway hyperreactivity/asthma	D'Alessandro et al. 1996
0.5	15 min	Change in nasal air resistance in rhinitic subjects (no change in nonrhinitic subjects); no effect on pulmonary peak flow, rhinorrhea, postnasal drip, or headache in either type of subject	Shusterman et al. 1998
0.5	8 h	Perception of odor, no discomfort, no effects, no changes in pulmonary function measurements for healthy individuals; some changes for atopic individual	Anglen 1981; Rotman et al. 1983
1.0	1 h	Statistically significant but modest changes in FEV_1 and R_{aw} for normal and asthmatic subjects	D'Alessandro et al. 1996
1.0	2 h	No noticeable effects	Joosting and Verberk 1974
1.0	4 h	Irritation for some sensations; no changes in pulmonary function measurements	Anglen 1981
1.0	4 h	Transient changes in pulmonary function measurements (airway resistance)	Rotman et al. 1983
1.0	8 h	Irritation (itchy eyes, runny nose, mild burning in throat); transient changes in pulmonary function measurements; atopic subject could not complete full 8-h exposure because of wheezing and shortness of breath	Anglen 1981; Rotman et al. 1983

2.0	15 min	Perception of odor; no significant irritation effects	Anglen 1981
2.0	30 min	Not significantly different from control group for irritant effects, irritancy indices	Anglen 1981
2.0	1 h	Itching or burning of throat, urge to cough at nuisance level	Anglen 1981
2.0	2 h	Very slight irritation of eyes, nose, and throat; no changes in pulmonary function	Joosting and Verberk 1974
2.0	2 h	No significant changes in pulmonary function	Anglen 1981
2.0	4 h	50% response of subjects to sensations characterized as nuisance: itching or burning of nose or throat, urge to cough, runny nose, general discomfort; transient changes in pulmonary function	Anglen 1981
2.0	8 h	Not immediately irritating, objectionable after several hours; increased mucous; transient changes in pulmonary function	Anglen 1981
4.0	2 h	Nuisance level of throat irritation, perceptible to nuisance level of nose irritation and cough	Joosting and Verberk 1974

[a]The Anglen (1981) and Joosting and Verberk (1974) studies were performed with healthy adults. Atopic individuals were included in the Shusterman et al. (1998) and Rotman et al. (1983) studies, and healthy subjects as well as asthmatic subjects were included in the D'Alessandro (1996) study.

[b]8-h studies were composed of two segments with a 30-min or 1-h break after 4 h.

In the first part of a two-part experiment, 31 un-acclimated male and female subjects (age range 20-32 years [y]) were exposed to chlorine at 0.0, 0.5, 1.0, or 2.0 ppm for 4 h or at 0.5 ppm or 1.0 ppm for 8 h (Anglen 1981). Not all subjects were exposed to all concentrations. Exposure days were randomly assigned, and the subjects did not know the test concentration, although the investigator did. In part two, eight nonsmoking males, ages 23-33 y, were exposed to concentrations of chlorine at 0.0, 0.5, or 1.0 ppm for 8 h. The 8-h sessions were broken into two 4-h sessions with a 30- or 60-min lunch break. A 15-min exercise period during each hour of exposure was designed to increase the average heart rate to 100 beats per minute. During the exposures, the subjects filled out a subjective questionnaire concerning 14 sensations ranging from smell to shortness of breath and using a scale of 0 (no sensation) to 5 (unbearable). Eye irritation was documented photographically; other signs of irritation were documented with pre- and post-exposure examinations by a physician. Pulmonary function tests (forced vital capacity [FVC] and forced expiratory volume at 1 second [FEV_1]) for each subject were measured before, during, and after exposures. The data were analyzed in terms of mean or median response, percent responding to greater than or equal to a set value, and responses to ranges or indices of irritation; the data were analyzed statistically where appropriate ($p > 0.167$, $p > 0.025$). Paired t-tests involving differences in values between pre-exposure and post-exposure times were used to analyze the pulmonary function measurements. Chlorine concentrations in the exposure chamber, measured by a variety of colorimetric and instrumental methods, were consistent.

At the tested concentrations, most of the subjects did not consider the sensations of smell or taste of chlorine unpleasant; therefore, those sensations were not included in the remainder of the analyses. In part one, the greatest number of subjects responded positively to the irritant sensation of itching or burning of the throat (described by the subjects as "feeling as if they had been talking for a long time"). Statistical differences were seen for that sensation at 1 and 2 ppm compared with controls; the level of response was ≥3 (nuisance or greater) for the 2-ppm concentration. A concentration of 0.5 ppm resulted in subjective irritation values between 1 (just perceptible) and 2 (distinctly perceptible) and produced no change in pulmonary functions. In part two, exposure at 1 ppm produced statistically significant changes in pulmonary function and increased subjective irritation at 8 h. No significant differences in pulmonary function measurements were seen at the end of a 4-h exposure at 1 ppm. Some differences in pulmonary function were seen at the end of 4 h at the 2-ppm concentration, but not at the end of 2 h. Most of the exposed subjects did not report sensations of nausea, headache, dizziness, or drowsiness at any concentration during the exposures. Male subjects were more sensitive to the irritant effects of chlorine than female subjects. The author concluded that exposure at 2 ppm for up to 30 min produced no significant increase in subjective irritation (severity of response not stated) over that seen during control exposures, and 2-h exposures at 2 ppm or 4-h exposures at 1 ppm produced no significant changes in pulmonary function (Anglen 1981).

In a follow-up study, Rotman et al. (1983) reported pulmonary function tests of eight male subjects, ages 19-33 y, exposed at 0.0, 0.5, and 1.0 ppm, like in the Anglen (1981) study. Air samples were collected eight times daily, and analysis for chlorine was accomplished using a modified NIOSH-recommended methyl orange method. The subjects were blind to the exposure concentration, but the study investigators were not. While in the chamber, each subject exercised for 15 min of each hour on an inclined treadmill or a by a simple step test at a rate that produced a heart rate of 100 beats per minute. Subjects exited the exposure chamber after 4 h to undergo the pulmonary function tests and then reentered the chambers. Comparisons of pre- and post-exposure pulmonary functions were made by paired t-tests between the percent change from baseline values obtained at analogous times after a sham exposure. Insignificant differences were observed with the sham versus the 0.5-ppm exposure. Compared with the sham changes from baseline, the changes in the 1.0-ppm exposure group were small, but statistically significant ($p < 0.05$); those changes were in FEV_1, peak expiratory flow rate (PEFR), forced expiratory flow rate at 50% and 25% vital capacity (FEF_{50} and FEF_{25}), total lung capacity (TLC), airway resistance (R_{aw}), and difference in nitrogen concentration between 750 milliliters (mL) and 1,250 mL of exhaled vital capacity (ΔN_2). At 8 h, changes were present in FVC, FEV_1, forced expired volume in 1 second as %FEV ($FEV_{1\%}$), PEFR, FEF_{50}, FEF_{25}, and R_{aw}. R_{aw} had the greatest response to chlorine exposure with an increase of 31% after 4 h of exposure at 1.0 ppm compared with increases of up to 6% during sham exposures. Most of those parameters had returned to pre-exposure values by the following day. However, a ninth subject whose pre-exposure lung parameters indicated obstructive airway disease (as defined by DuBois et al. [1971]) did not complete the full 8 h of exposure at 1.0 ppm because of shortness of breath and wheezing; his values were not included in the statistical analysis. The atopic individual experienced changes in several pulmonary function parameters after exposure to chlorine at 0.5 ppm. The greatest change for this individual was in R_{aw}, which increased by 40% over the pre-exposure value after 4 h of exposure at 0.5 ppm and by 33% over the pre-exposure value after 8 h of exposure at 0.5 ppm. Changes in R_{aw} in the healthy subjects were 5% and 15% for the respective time periods. Following the 4-h exposure at 1.0 ppm, R_{aw} increased from a pre-exposure value of 3.3 centimeters (cm) of water per liter per second to 14.4 cm of water per liter per second in the atopic individual. The authors concluded that at the 1-ppm concentration, serious subjective symptoms of irritation were not produced in healthy adults, but transient altered pulmonary function was observed.

The authors discussed the implications of changing baselines and significant changes over time with sham exposures, the latter with respect to diurnal variation in pulmonary function. Baselines differed on different days of the tests (e.g., the pre-exposure baselines for total lung capacities [TLCs] were 7.09 L and 6.55 L on two different days, which is significantly different [$p < 0.05$]; the TLC decreased after the sham exposure, but increased after the 0.5-ppm exposure). However, differences in parameters between baseline and post-exposure for sham exposures were fewer in number and generally smaller in amount than for the 1 ppm treatment. It should be noted that in normal subjects, several pulmonary function tests (FVC, FEV_1, and $FEF_{25-75\%}$) may have daily changes of 5% to 13% and week to week changes of 11% to 21% (EPA 1994).

D'Alessandro et al. (1996) exposed 12 male and female volunteers, ages 18-50 y, to chlorine for 1 h. Five of the subjects were without airway hyper-reactivity (defined by baseline methacholine hyper-responsiveness) and seven were diagnosed with airway hyper-reactivity. Five of the seven with airway hyper-reactivity had clinical histories of asthma (one was being treated regularly with corticosteroids). Ten subjects, five normal and five hyper-reactive (three of which had asthma), were exposed to a concentration of chlorine at 1.0 ppm. Five of the subjects with asthma were exposed at 0.4 ppm. They were not blinded to the exposure status. The subjects were exposed to chlorine by mask while in the sitting position; there were no air exposures. Chlorine was measured with a chlorine analyzer. The following pulmonary function parameters were measured or calculated immediately following and 24 h after exposure: FEV_1, TLC, carbon monoxide diffusing capacity (D_{CO}), R_{aw}, and $FEF_{25-75\%}$. After asthmatic subjects were exposed at 0.4 ppm, there were no statistically significant changes in any parameters, including FEV_1 and R_{aw}, either immediately following or 24 h after exposure. Immediately following the exposure at 1.0 ppm, there were statistically significant changes in FEV_1 and R_{aw} for both normal and hyper-reactive subjects compared with baseline values. Hyper-reactive subjects showed a greater relative decrease in FEV_1 (16% compared with

4% for normal subjects) and a greater relative increase in R_{aw} (108% compared with 39% for the normal subjects). Although one hyper-responsive subjects' FEV_1 fell by 1,200 mL, and the R_{aw} more than tripled, the mean changes were considered modest by the authors. The hyper-responsive subject with the greatest increase in R_{aw} following exposure at 1.0 ppm showed virtually no change following exposure at 0.4 ppm. Two subjects characterized as hyper-responsive experienced undefined respiratory symptoms following exposure at 1.0 ppm. For all subjects, most values were close to baseline by 24 h post-exposure. The fact that none of the subjects found the odor of chlorine appreciable at either concentration is interesting.

In a single-blind crossover study, Shusterman et al. (1998) measured nasal air resistance via active posterior rhinomanometry in eight subjects with seasonal allergic rhinitis and eight nonrhinitic subjects. Measurements were made before, immediately after, and 15 min after a 15-min exposure to either filtered air or chlorine at 0.5 ppm in filtered air administered through a nasal mask in a climate-controlled chamber. Each subject served as his or her control, and subjects were free of medications for at least 24 h prior to testing. Subjects were between 18 and 40 y of age. The mean percent change in nasal air resistence from baseline to immediately after exposure was +24% in the subjects with allergic rhinitis and +3% in the nonrhinitic group. The mean percent change from baseline to 15 min after exposure was +21% in the subjects with allergic rhinitis and -1% in the nonrhinitic subjects. Differences between groups were significant ($p < 0.05$) for both post-exposure times. Rhinitic subjects reported greater exposure-related increases in odor intensity, nasal irritation, and nasal congestion than did nonrhinitic subjects, but the relationship between subjective and objective nasal congestion was weak. No significant exposure-related changes were observed for rhinorrhea, postnasal drip, or headache. Pulmonary peak flow was also obtained before and after exposure and none of the subjects exhibited clinically significant changes in peak flow (decreases of ≥10% of baseline), nor did they complain of cough, wheezing, or chest tightness during chlorine exposure days.

Joosting and Verberk (1974) exposed eight subjects, ages 28-52, at 0.5-4 ppm for 2 h. The subjects were all members of a Dutch subcommittee on toxicology. Subjective reactions were noted every 15 min. Subjects exited the chambers every 15 min to perform spirometry tests. Concentrations of chlorine at 0.5 ppm and 1.0 ppm did not produce noticeable effects in 2 h; 2 ppm produced very slight eye, nose, and throat irritation; and 4 ppm resulted in a distinctly perceptible to offensive level of irritation of the nose and throat and desire to cough. The highest score was for irritation of the throat, for which the average response was nuisance. No sensory irritation scores reached unbearable. No effects on lung function (vital capacity [VC], FEV, and forced inspiratory volume [FIV]) occurred at the lower concentrations; effects on lung function were not reported in the 4-ppm exposure group because only 2-3 subjects completed the exposure.

Older studies were located in the literature; those have been reviewed and critiqued by NIOSH (1976) and OSHA (1989). In several older studies in which measurement techniques and/or ranges of measured values were not given, concentrations as low as 0.027 ppm produced slight sensory effects in humans and concentrations at 0.5-4.0 ppm produced sensory irritation (NIOSH 1976). For example, Rupp and Henschler (1967) reported that human volunteers experienced burning of the eyes after exposure at 0.5 ppm for 15 min. In a separate test, the subjects reported respiratory irritation during exposure at 0.5 ppm and discomfort during exposure at 1 ppm. The study has been criticized for its lack of controls as well as the possible presence of confounding chemicals (OSHA 1989). However, the results of the Rupp and Henschler (1967) study that indicated some irritation at 0.5 ppm and 1 ppm are not all that different from the results of Anglen (1981), Rotman et al. (1983), and D'Alessandro et al. (1996).

2.2.2. Epidemiologic Studies

Few epidemiologic studies document average chlorine exposure concentrations over long periods of time; concentration variations over time; or exposure durations to various concentrations. Interpretation of results is often complicated by unknown previous exposures to chlorine, exposures to other chemicals, and smoking habits. NIOSH (1976) discussed available epidemiological studies conducted prior to 1976. In most of the cited studies, work-room concentrations averaged <1 ppm. The report noted the difficulty in correlating exposures to effects.

Patil et al. (1970) compared the health of 382 workers in 25 chlorine production plants in the United States and Canada with that of unexposed workers in the same plants. All subjects were male, between the ages of 19 and 69. Time-weighted average exposures to chlorine ranged from 0.006 ppm to 1.42 ppm, with a mean of 0.146 ppm; almost all workers were exposed to <1 ppm. The average number of exposure years was 10.9. There were no statistically significant ($p < 0.05$) signs or symptoms on a dose-response relation basis in chest x-rays, electrocardiograms, or pulmonary function tests. Nor were there dose-response relationships with cough, sputum production, frequency of colds, dyspnea, palpitation, chest pain, fatigue, tremors, gastrointestinal problems, dermatitis, or hematologic parameters. Subjective complaints of tooth decay were dose-related, but that complaint was not borne out by physical examination.

A study of respiratory effects in 52 Italian electrolytic cell workers with an average exposure to chlorine of 0.298 ± 0.181 ppm was undertaken by Capodaglio et al. (1970, as cited in ACGIH 1995, 1996). Of five respiratory function tests (FEV_1, VC, D_{CO}, residual volume, and helium concentration gradient in a single breath during washout), only carbon monoxide diffusing capacity showed a slight but significant difference; however, cigarette smoking may have contributed to that observation.

Mortality and morbidity (respiratory symptoms, disease, and functions) of workers exposed to chlorine in the pulp and paper industry over a 10-y period (1963-1973) were similar to those of the general white male population (Ferris et al. 1979). Mean and maximum exposure concentrations in the pulp mill were to trace amounts (< 0.0005 ppm). In an earlier study (Ferris et al. 1964), exposures in the pulp mill were more variable (mean, 7.4 ppm; range up to 64 ppm) and slight adverse effects on respiratory ailments were found. Exposures were also to chlorine dioxide.

Between 1984 and 1989, a prospective study was conducted on the effects of chlorine exposure on workers in a chlorine manufacturing plant (Kusch 1994). Chlorine exposures and pulmonary function tests (FVC, FEV_1, and $FEF_{25\text{-}50\%}$), taken over a period of 5 y were compared with a control group. The average exposure in the control group was 0.058 ppm, and the average exposure in the workers was 0.092 ppm. There were no measurable effects on pulmonary function related to chlorine exposure.

2.2.3. Accidents

An accident at a chemical plant in India resulted in the exposure of 88 workers as well as police and fire-fighting personnel to a measured concentration of 66 ppm for an unspecified amount of time (Shroff et al. 1988). No further details on the exposure concentration or duration were given. The workers, ages 21-60 y, were admitted to the hospital within an hour of exposure with symptoms of dyspnea, coughing, irritation of the throat and eyes, headache, giddiness, chest pain, and abdominal discomfort. Examinations revealed hilar congestion, bronchial vasculature, respiratory incapacitation at the PFT (undefined); bronchoscopy revealed tracheobronchial congestion, chronic bronchitis, scattered hemorrhages, and bronchial erosion. Bronchial smears of 28 patients on day 5 after the accident showed basal-cell and goblet-cell hyperplasia; acute inflammation; and chromatolysis of columnar epithelial cells, multinucleated syncytial respiratory epithelial cells with degenerating cilia, and nonpigmented alveolar macrophages. In some patients these effects progressed to bronchopneumonia, epithelial regeneration, and repair by fibrosis by day 25 post-exposure.

During an industrial accident, a group of workers was presumably exposed to concentrations up to 30 ppm (based on symptoms—actual exposure concentration and duration not known) (Abhyankar et al. 1989). Initial symptoms included watering eyes, sneezing, cough, sputum, retrostenal burning, dyspnea, apprehension, and vomiting. All patients were asymptomatic by 2 weeks (wk) post-exposure; at 6 months (mo) post-exposure, all spirometry tests (FVC, FEV_1) were within the normal range. Those patients known to have a pre-existing lung condition did not show any additional evidence of lung damage.

Incidents of acute poisonings at indoor swimming pools have been reported (Decker 1988), but air concentrations during those incidents are unknown. In a study in Spain, the mean air concentration measured during five nonconsecutive days in four enclosed swimming pools was 0.42 ± 0.24 milligrams per cubic meter (mg/m^3) (0.14 ± 0.08 ppm) (Drobnic et al. 1996). The samples were taken at <10 cm above the water, the breathing zone of swimmers.

Acute exposure to chlorine/chloramine gas occurs often among the general public through the mixing of domestic home cleaners (Mrvos et al. 1993); swimming pool chlorinator tablets (Wood et al. 1987); and intentional self administration (Rafferty 1980). In the case of home exposures, a review of 216 cases reported to a Regional Poison Information Center showed that symptoms, primarily cough with resulting shortness of breath, resolved within 1 to 6 h without medical intervention. There was no information on exposure concentrations (Mrvos et al. 1993).

2.3. Developmental and Reproductive Effects

No studies on developmental and reproductive effects in humans were located.

2.4. Genotoxicity

No data concerning the genotoxicity of chlorine in humans via inhalation exposures were identified in the available literature. When chlorine (sodium hypochlorite) at concentrations of 20 ppm and above was added to cultures of human lymphocytes, chromosomal aberrations (breaks and rearrangements) and endomitotic figures were observed (Mickey and Holden 1971).

2.5. Carcinogenicity

No increase in neoplasms was reported in an epidemiology study of workers engaged in the production of chlorine (Patil et al. 1970). The range of time-weighted exposures to chlorine was 0.006-1.42 ppm.

2.6. Summary

Anglen (1981) conducted a study on 31 male and female subjects in which slight but statistically significant changes in pulmonary function and subjective irritation resulted from exposure to chlorine at 1 ppm for 8 h (two 4-h sessions). Subjective sensory irritation (itching or burning of the throat) was described as "just perceptible" or "distinctly perceptible." A 30-min exposure at 2 ppm produced no increase in subjective irritation. An 8-h exposure at 0.5 ppm produced no changes in lung function and no significant sensory irritation. A 4-h exposure at 1 ppm produced no changes in pulmonary function tests, but an 8-h exposure at 1 ppm produced slight declines in some pulmonary function tests. Most of these findings were confirmed in a study by Rotman et al. (1983) in which eight healthy volunteers were exposed at 0.5 ppm or 1.0 ppm for an interrupted 8 h. Transient but statistically significant declines in six of 15 pulmonary function tests were associated with exposure at 1 ppm for 4 or 8 h but not with exposure at 0.5 ppm for 4 or 8 h. These studies reported that there was no effect of chlorine exposure on carbon monoxide diffusing capacity, thus indicating there is no significant pulmonary edema from the exposures. However, an atopic subject in the Rotman et al. study (1983) suffered an asthma-like attack resulting from exposure to chlorine at 1 ppm; that subject withstood exposure, testing, and exercise for the first 4 h, but exited the exposure chamber "before the full 8-h exposure to 1 ppm." Changes in his pulmonary function measurements were greater than those of the other test subjects. The atopic subject completed the interrupted 8-h exposure at 0.5 ppm.

The Anglen and Rotman studies were supported by two additional studies. In the first study (Angelen 1981), subjects with airway hyper-reactivity, including asthmatic subjects, showed no significant changes in FEV_1 or R_{aw} following a 1-h exposure at 0.4 ppm. Exposure of both normal and hyper-reactive subjects at 1.0 ppm significantly decreased FEV_1 and significantly increased R_{aw} in both sets of subjects; the hyper-reactive subjects showed a significantly greater response than normal subjects. However, the mean changes were considered modest by the authors. In the second study, subjective sensory irritation of healthy individuals reached "nuisance" level at an exposure concentration of 4 ppm for 2 h (Joosting and Verberk 1974). This study showed that time, as well as concentration, was a factor in subjective response to chlorine inhalation.

No useful exposure data could be derived from epidemiology studies or human exposures to accidental releases of chlorine. No studies were located on developmental and reproductive effects. In an in vitro study, chlorine was genotoxic at 20 ppm.

3. ANIMAL TOXICITY DATA

3.1 Acute Lethality

A summary of the acute lethality data is presented in Table 1-4. Some of those studies were reviewed by Withers and Lees (1985a). According to Withers and Lees (1985a), many of the older studies had deficiencies in gas analysis methods and exposure conditions. More recent studies contradict the results in the older studies (e.g., the 3-h LC_{50} of 10 ppm in mice reported by Schlagbauer and Henschler [1967, as cited in AIHA 1988] is contradicted by the nonlethal 6-h exposure at 9.3 ppm for 5 d reported in the more recent study by Buckley et al. [1984]). Nevertheless, some of the older studies (Lipton and Rotariu 1941, as cited in Withers and Lees 1985a; Silver et al. 1942, as cited in Withers and Lees 1985a; Schlagbauer and Henschler 1967, as cited in AIHA 1988) are cited in Table 1-4 for comparison purposes. More recent studies are discussed below.

3.1.1. Dogs

In an early study, Underhill (1920) reported on mortality in dogs following 30-min exposures to a range of concentrations (50-2,000 ppm). A total of 112 male and female dogs of several breeds were used. Acute mor-

TABLE 1-4 Summary of Acute Lethal Inhalation Data in Animals

Species	Concentration (ppm)	Exposure Time	Effect[a]	Reference
Dog	650	30 min	LC_{50}	Underhill 1920; Withers and Lees 1985a
Rat	5,500 2,841	5 min 5 min	LC_{50} No deaths	Zwart and Woutersen 1988
Rat	1,946	10 min	LC_{50}	Zwart and Woutersen 1988
Rat	700 547	30 min 30 min	LC_{50} No deaths	Zwart and Woutersen 1988
Rat	1,000	53 min	LC_{50}	Weedon et al. 1940
Rat	455	1 h	LC_{50}	Zwart and Woutersen 1988
Rat	288[b] 322	1 h 1 h	LC_{01} No deaths	Zwart and Woutersen 1988
Rat	293[c] 213	1 h 1 h	LC_{50} No deaths	Back et al. 1972; MacEwen and Vernot 1972; Vernot et al. 1977
Rat	250	7.3 h	LC_{50}	Weedon et al. 1940
Rat	63	>16 h	LC_{50}	Weedon et al. 1940
Mouse	290	6 min	No deaths	Bitron and Aharonson 1978
Mouse	1,057 754	10 min 10 min	LC_{50} No deaths	Zwart and Woutersen 1988
Mouse	676	10 min	LC_{50}	Silver et al. 1942[d]
Mouse	628	10 min	LC_{50}	Lipton and Rotariu 1941[d]
Mouse	549	10 min	25-45% mortality	Silver et al. 1942[d]
Mouse	380	10 min	10% mortality	Silver et al. 1942[d]
Mouse	302	10 min	LC_{50}	Alarie 1980
Mouse	290	11 min	LC_{50}	Bitron and Aharonson 1978

(Continued)

TABLE 1-4 Continued

Species	Concentration (ppm)	Exposure Time	Effect[a]	Reference
Mouse	290	15 min	80% mortality	Bitron and Aharonson 1978
Mouse	290	25 min	100% mortality	Bitron and Aharonson 1978
Mouse	1,000	28 min	LC_{50}	Weedon et al. 1940
Mouse	504	30 min	LC_{50}	Zwart and Woutersen 1988
Mouse	127 55	30 min 30 min	LC_{50} No deaths	Schlagbauer and Henschler 1967[e]
Mouse	170	55 min	LC_{50}	Bitron and Aharonson 1978
Mouse	137[c]	1 h	LC_{50}	Back et al. 1972; MacEwen and Vernot 1972; Vernot et al. 1977
Mouse	250 200 150	1 h 1 h 1 h	LC_{80} LC_{01} No deaths	O'Neil 1991
Mouse	170	2 h	80% mortality	Bitron and Aharonson 1978
Mouse	10	3 h	80% mortality[f]	Schlagbauer and Henschler 1967[e]
Mouse	250	7.3 h	LC_{50}	Weedon et al. 1940
Mouse	63	>16 h	LC_{50}	Weedon et al. 1940
Rabbit	500	30 min	100% mortality	Barrow and Smith 1975

[a]LC_{50} and LC_{100} values were obtained immediately after exposure (Weedon et al. 1940), 3 h post-exposure (Alarie 1980), 10 d post-exposure (Silver et al. 1942), 14 d post-exposure (Back et al. 1972; MacEwen and Vernot 1972; Vernot et al. 1977; Zwart and Woutersen 1988), and 30 d post-exposure (Bitron and Aharonson 1978).
[b]Calculated by Zwart and Woutersen (1988) using probit analysis; note this value is lower than the concentration resulting in no deaths.
[c]Authors report a 20-30% loss of chlorine in the exposure chambers; the concentrations given are measured concentrations.
[d]As cited in Withers and Lees 1985a.
[e]As cited in AIHA 1988.
[f]These results conflict with results of other studies.

talities, defined as deaths within 3 d of the exposure, were 0%, 6%, 20%, 43%, 50%, 87%, and 92% at concentration ranges of 50-250 ppm, 400-500 ppm, 500-600 ppm, 600-700 ppm, 700-800 ppm, 800-900 ppm, and 900-2,000 ppm, respectively. However, some delayed deaths, occurring as the result of bronchopneumonia following subsidence of acute pulmonary edema, resulted in all groups; one of nine suffered a delayed death (time not given) at the 50-250 ppm range. Withers and Lees (1985a) analyzed these data using the method of Litchfield and Wilcoxon (1949) and calculated an LC_{50} of 650 ppm. That value is based on the concentration in ppm at 25°C. Underhill (1920) initially made the conversion from mg/m^3 to ppm based on a temperature of 0°C. Dogs exposed to chlorine became excited and displayed signs of respiratory irritation; those signs were followed by labored breathing (Underhill 1920).

3.1.2. Rats

Back et al. (1972), MacEwen and Vernot (1972), and Vernot et al. (1977) reported the same 1-h LC_{50} of 293 ppm (95% confidence limits, 260-329 ppm) for Sprague-Dawley rats. MacEwen and Vernot (1972) noted a 20-30% loss of chlorine in the exposure chambers, probably due to condensation on the walls. Therefore, the concentrations given are measured concentrations. Rats experienced immediate eye and nose irritation followed by lacrimation, rhinorrhea, and gasping after 1 h of exposure at all tested concentrations (213, 268, 338, and 427 ppm). Rats surviving the 213 ppm and 268-ppm exposures gained less weight than the control group during the 14-d post-exposure period. No deaths occurred in rats exposed at 213 ppm for 1 h. Weedon et al. (1940) exposed groups of eight rats to 63, 240, or 1,000 ppm for 16 h or until death. Times to 50% mortality were >16 h, 7.3 h, and 53 min, respectively.

Zwart and Woutersen (1988) exposed specific pathogen free (SPF) Wistar-derived rats to chlorine at 322-5,793 ppm for exposure durations of 5 min to 1 h to calculate LC_{50} values. Observations were made over a 14-d period after which the animals were sacrificed and histologic examinations were made; additional groups of rats were exposed and examined 2 d after exposure. Rats exposed at the highest concentrations during the 30- and 60-min exposures showed signs of restlessness, eye and nasal irritation, labored breathing, and reduced respiratory rate. Mortalities occurred during exposure as well as within the first week of the observation period. Increased lung weights were positively correlated with higher concentrations and longer exposure durations. At the high concentrations, 5,793 ppm for 5 min and 2,248 ppm for 10 min, effects were observed in the nose, larynx, and trachea; at those and the lower concentrations, lung lesions, including focal aggregates of mononuclear inflammatory cells, increased septal cellularity, squamous metaplasia of bronchiolar epithelium, and edema, were observed in one or more animals. Hyperplasia of the larynx and trachea observed at 2 d post-exposure was resolved by 14 d post-exposure in surviving rats. In rats in which minute-volume was measured, death occurred in several animals following a reduction in minute-volume to ≤39% of the pre-exposure level. According to the authors, the 1-h LC_{01} of 288 ppm (95% confidence interval, 222-345 ppm), estimated by probit analysis, appeared to correspond with the onset of irreversible lung damage. The breathing pattern during the exposures changed from regular inspiration directly followed by a regular expiration to rapid shallow breathing that lasted less than a minute. That was followed by maximal inhalation directly after expiration and a long post-inspiratory pause. No deaths occurred at 2,841 ppm for 5 min, 547 ppm for 30 min, or 322 ppm for 60 min.

In a subchronic study, three of 10 female F-344 rats exposed at 9 ppm for 6 h/d, 5 d/wk died by the 30th exposure (Barrow et al. 1979). In another study, groups of eight 30-wk-old male and female SPF rats were exposed to chlorine at approximately 117 ppm for 3 h/d, 7 d/wk until half of the animals of each gender died (Bell and Elmes 1965). Total exposure time to 50% mortality was 29 h for males and 32 h for females. When the response of conventional rats was compared with that of SPF rats, the response was more severe in the conventional rats, who exhibited proliferation of goblet cells and increased mucus, emphysema, and polymorphonuclear cells in the lungs.

3.1.3. Mice

Back et al. (1972), MacEwen and Vernot (1972), and Vernot et al. (1977) reported a 1-h LC_{50} of 137 (95% confidence limits, 119-159 ppm). Weedon et al. (1940) exposed groups of four mice to 63, 240, or 1,000 ppm for 16 h or until death. Times to 50% mortality were >16 h, 7.3 h, and 28 min, respectively.

Bitron and Aharonson (1978) exposed male albino mice to 170 ppm and 290 ppm for several exposure times and calculated 50% mortality as a function of exposure time (Lt_{50}). Mice were restrained during the exposures. Observations were made over a 30-d period. Lt_{50} for the 170-ppm and 290-ppm exposures were 55 min and 11 min, respectively. The results of this work were unusual in that many of the deaths were delayed, occurring during the second week of the observation period, rather than during and immediately following exposure. No mice died within a 30-d observation period following exposure at 290 ppm for 6 min. No deaths occurred in BALB/c mice exposed at 50, 100, or 150 ppm for 1 h (O'Neil 1991). Mice were observed for at least 5 d post-exposure.

Zwart and Woutersen (1988) exposed Swiss mice to chlorine at 579-1,654 ppm for 10 min and 458-645 ppm for 30 min to calculate LC_{50} values. Mortality observations were made over a 14-d period. Nearly one-third of the mice died during the second week post-exposure, indicating to the authors that the deaths may have been due to secondary infection. Increased lung weights were positively correlated with higher concentrations and longer exposure durations. No deaths occurred at a concentration of chlorine at 754 ppm for 10 min.

Alarie (1980) reported that the 10-min LC_{50} of male Swiss-Webster mice decreased from 302 ppm in uncannulated mice to 131 ppm when chlorine was delivered directly to the trachea via cannulation.

As part of an experiment on immune response, groups of 10 BALB/c mice were exposed at 50, 100, 150, 200, or 250 ppm for 1 h (O'Neil 1991). Mortality occurred at 200 ppm (two mice, 4-5 d post-exposure) and at 250 ppm (8/10 mice).

3.2. Nonlethal Toxicity

Data on effects following exposures to nonlethal concentrations of chlorine are available for the monkey, rat, mouse, guinea pig, and rabbit. Studies utilizing acute exposure durations are summarized in Table 1-5.

3.2.1. Nonhuman Primates

No studies on single acute exposures were located. Klonne et al. (1987) exposed Rhesus monkeys to chlorine at 0, 0.1, 0.5, or 2.3 ppm for 6 h/d, 5 d/wk for 1 y. In the group exposed at 2.3 ppm, ocular irritation as well as treatment-related histopathologic changes limited to the nasal passages and trachea were observed at 1 y. Those lesions, consisting of focal, epithelial hyperplasia with loss of cilia and decreased numbers of goblet cells, were considered mild in the group exposed at 2.3 ppm and were not present in all animals. Lesions in the lower exposure groups were minimal. No statistically significant differences were observed between control and exposure groups for pulmonary diffusing capacity of carbon monoxide or distribution of ventilation values (number of breaths to 1% N_2).

3.2.2. Rats

Demnati et al. (1995) exposed groups of four male Sprague-Dawley rats (nose-only) to chlorine at 0, 50, 100, 200, 500, or 1,500 ppm for 2-10 min in order to study effects on airway mucosa and lung parenchyma. Histologic examinations were performed at 1, 3, 6, 12, 24, and 72 h after exposure. Exposures to concentrations of ≤500 ppm did not induce significant histologic changes. Lungs from control rats and from rats exposed at 50-100 ppm for 2 min were normal within 72 h; at concentrations of 200 ppm and 500 ppm for 2-5 min, there was only slight perivascular edema in all exposed rats. Exposure at 1,500 ppm for 2 min produced only slight effects, including mild perivascular edema and occasional small clusters of polymorphonuclear leukocytes in the mucosa of large airway. The 10-min exposure at 1,500 ppm caused significant changes that varied with time after exposure—airspace and interstitial edema associated with bronchial epithelial sloughing at 1 h, decreased edema and the appearance of mucosal polymorphonuclear leucocytes at 6-24 h, and epithelial regeneration as evidenced by hyperplasia and goblet cell metaplasia at 72 h. No deaths were reported.

Exposure at 25 ppm lowered the respiratory rate by 50% (RD_{50}) in F-344 rats, presumably during a 10-min test (Barrow and Steinhagen 1982);during exposure for 6 h, the RD_{50} was 10.9 ppm (Chang and Barrow 1984). Groups of 9-10 male F-344 rats were exposed to a concentration of chlorine at 9.1 ppm (the RD_{50} of mice) for 1, 3, or 5 d (6 h/d) and examined for respiratory tract pathology (Jiang et al. 1983). Sacrifice took place immediately after exposure. In all animals, lesions were present in the nasal passages with less severe changes in the nasopharynx, larynx, trachea, and lungs. Lesions in the nasal passages involved epithelial degeneration with epithelial cell exfoliation, erosion, and ulceration (respiratory epithelium) and extensive epithelial erosion and ulceration (olfactory epithelium of the dorsal meatus). Electron microscopy examination revealed loss of respiratory and olfactory cilia and cellular exfoliation of the naso- and maxilloturbinates.

TABLE 1-5 Summary of Sublethal Effects in Laboratory Animals

Species	Concentration (ppm)	Exposure Time	Effect[a,b]	Reference
Rat	1,500	2 min	Mild perivascular edema of lung, leucocytic infiltration	Demnati et al. 1995
Rat	200, 500	2, 5 min	Slight perivascular edema of lung	Demnati et al. 1995
Rat	50, 100	2 min	No effect	Demnati et al. 1995
Rat	1,500	10 min	Epithelial hyperplasia, goblet cell metaplasia of lung	Demnati et al. 1995
Rat	25	10 min	RD_{50}	Barrow and Steinhagen 1982
Rat	10.9	6 h	RD_{50}	Chang and Barrow 1984
Rat	9.1	6 h	Lesions in nasal passages; less severe changes in nasopharynx, larynx, trachea, and lungs	Jiang et al. 1983
Mouse	9.3	10 min	RD_{50}	Barrow et al. 1977
Mouse	3.5	1 h	RD_{50}	Gagnaire et al. 1994
Mouse	9.1	6 h	Lesions in nasal passages; less severe changes in nasopharynx, larynx, trachea, and lungs	Jiang et al. 1983
Rabbit	50	30 min	No gross or microscopic lung changes	Barrow and Smith 1975
Rabbit	100, 200	30 min	Initial changes in lung function; hemorrhage, pneumonitis, bronchitis; recovery at 60 d except pulmonary compliance	Barrow and Smith 1975

[a]Observed immediately after exposure (Jiang et al. 1983), 72 h post-exposure (Demnati et al. 1995), 5 d post-exposure (O'Neil 1991), 14 d post-exposure (MacEwen and Vernot 1972; Barrow and Smith 1975; Zwart and Woutersen 1988), 30 d post-exposure (Bitron and Aharonson 1978).

[b]The RD_{50} test is usually a 10-min test.

When male and female F-344 rats were exposed to chlorine at 0, 1, 3, or 9 ppm for 6 h/d, 5 d/wk for 6 wk, effects were observed in the upper and lower respiratory tract (Barrow et al. 1979). Lesions in male and female rats exposed at 9 ppm included widespread inflammation throughout the respiratory tract with hyperplasia and hypertrophy of epithelial cells of the respiratory bronchioles, alveolar ducts, and alveoli. Hepatocellular cytoplasmic vacuolation was observed in both genders exposed at 3 ppm or 9 ppm, and renal tubule effects were observed in male rats exposed at 9 ppm. Effects observed in animals exposed at 1 ppm or 3 ppm were much less severe than those observed at 9 ppm. In addition, decreased body weight gains were observed in females at all exposure concentrations and in males at 3 ppm and 9 ppm.

Groups of 14-wk-old male and female SPF rats were exposed to chlorine at approximately 40 ppm for 3 h/d for a total of 42 h (Bell and Elmes 1965). No deaths occurred. Signs and symptoms of exposure included coughing, sneezing, and runny and blood-stained noses after 3 h. Histologic examinations of the lungs revealed recovery by 14 d post-exposure. Exposure of conventional rats for 1 h daily to 14-18 ppm, for a total of 24 exposure hours in 4 wk, also resulted in no mortality (Elmes and Bell 1963). The authors considered the chlorine concentrations overestimates, because the exposed rats huddled together in the exposure cage.

Groups of 70 male and 70 female F-344 rats were exposed to chlorine gas at 0, 0.4, 1.0, or 2.5 ppm for 6 h/d, 5 d/wk (males) or 3 alternate d/wk (females) for 2 y (CIIT 1993; Wolf et al. 1995). Concentration-dependent lesions confined to the nasal passages were observed in all animals. These lesions were most severe in the anterior nasal cavity and included respiratory and olfactory epithelial degeneration, septal fenestration, mucosal inflammation, respiratory epithelial hyperplasia, squamous metaplasia and goblet cell hypertrophy and hyperplasia, and secretory metaplasia of the transitional epithelium of the lateral meatus. Body weights were depressed compared with controls, but no early deaths occurred.

3.2.3. Mice

Groups of four male Swiss-Webster mice were exposed to chlorine concentrations at 0.7-38.4 ppm for 10 min (Barrow et al. 1977). The RD_{50} was 9.3 ppm. The RD_{50} of male OF_1 mice was calculated at 3.5 ppm by Gagnaire et al. (1994). Although that exposure was for 60 min, the decrease occurred by 10 min. The protocol of ASTM (1991) was followed by Gagnaire et al., but the OF strain is not the strain of mice suggested for use in measuring sensory irritation. The ASTM (1991) RD_{50} test calls for male Swiss-Webster mice and a 10-min exposure period.

In a follow-up study to that of Barrow et al. (1977), Buckley et al. (1984) exposed male Swiss-Webster mice to the RD_{50} (9.3 ppm) for 6 h/d for 5 d. Half of each group was necropsied immediately after the last exposure and the other half was necropsied 72 h post-exposure. No deaths were reported. Lesions in both the anterior respiratory epithelium adjacent to the dorsal meatus and in the respiratory epithelium included exfoliation, inflammation erosion, ulceration, and necrosis. Chlorine reached the lower respiratory tract, as indicated by tracheal lesions and terminal bronchiolitis, with occlusion of the affected bronchioles by serocellular exudate. Recovery was minimal to moderate after 72 h.

Groups of 9-10 male Swiss-Webster mice were exposed at 9.1 ppm, the approximate RD_{50} for 1, 3, or 5 d (6 h/d) and examined for respiratory tract pathology (Jiang et al. 1983). Sacrifice took place immediately after exposure. Respiratory tract lesions were similar to those of the rat described above (lesions in the nasal passages with less severe changes in the nasopharynx, larynx, trachea, and lungs).

Groups of 70 male and 70 female $B6C3F_1$ mice were exposed to chlorine gas at 0, 0.4, 1.0, or 2.5 ppm for 6 h/d, 5 d/wk for 2 y (CIIT 1993; Wolf et al. 1995). Concentration-dependent lesions confined to the nasal passages were observed in all animals. These lesions were most severe in the anterior nasal cavity and included respiratory and olfactory epithelial degeneration, septal fenestration, mucosal inflammation, respiratory epithelial hyperplasia, squamous metaplasia and goblet cell hypertrophy and hyperplasia, and secretory metaplasia of the transitional epithelium of the lateral meatus. Body weights were depressed (males, all exposures; females, 2.5 ppm) compared with controls, but no early deaths occurred.

3.2.4. Guinea pigs

Arlong et al. (1940, as cited in NIOSH 1976) exposed guinea pigs to 1.7 ppm for 5 h daily over 87 d. No deaths were reported during the 300-d observation period. No other details of the study were available.

3.2.5. Rabbits

Groups of two male and two female rabbits were exposed at 0, 50, 100, or 200 ppm for 30 min and tested for lung changes as measured by volume-pressure relationships and inspiratory-expiratory flow rate at times from 30 min to 60 d post-exposure (Barrow and Smith 1975). Rabbits exposed at 50 ppm showed no changes at any time periods. Recovery of flow rate ratios occurred by 14 d post-exposure in the 100-ppm groups and by 60 d post-exposure in the 200-ppm group. Pulmonary compliance in the 100-ppm and 200-ppm groups did not return to control levels within 60 d. Examinations of the lungs revealed hemorrhages, pneumonitis and anatomic emphysema in the 100-ppm and 200-ppm groups at 3 and 14 d post-exposure; those changes were not present at 60 d post-exposure.

3.3. Developmental and Reproductive Effects

No studies addressing developmental or reproductive effects following inhalation exposure to chlorine were located. However, because effects on development and reproduction would be systemic, due to circulating chlorine, the effects of oral administration of chlorine may have bearing on the chlorine hazard assessment. Those data, reviewed by EPA (1996) and AIHA (1988), demonstrated no or insufficient evidence of reproductive or developmental toxicity.

Groups of 10 male B6C3F$_1$ mice were dosed by oral gavage with 1 mL of test solution containing OCl^- or HOCl at 40, 100, or 200 mg/L/d for 5 d (Meier et al. 1985). Animals were sacrificed at 1, 3, or 5 wk after the last treatment, and the caudae epididymides were examined for sperm head abnormalities. No abnormalities were observed at 1 and 5 wk post-treatment in the groups treated with OCl^- or at any time in the groups treated with HOCl. A small but statistically significance difference compared with controls was observed in the groups administered OCl^- at all dose levels at 3 wk. These results do not clearly indicate an effect on fertility.

Druckrey (1968) administered highly chlorinated drinking water (100 mg/L) to seven consecutive generations of BD II rats. The average daily dose was estimated at 10 mg/kg/d. No treatment-related effects were observed on any generation. Carlton et al. (1986) administered chlorine in deionized water by gavage male and female Long Evans rats at doses of 1.0, 2.0, or 5.0 mg/kg/d to for 66-76 d. Dosing began prior to mating and continued during gestation and lactation. Groups of offspring were necropsied at 21 d after birth or at 40 d of age; the latter group was dosed following weaning. No statistically significant differences were observed between the control and treated rats in litter survival, litter size, or pup weight.

HOCl, formed by bubbling chlorine gas through water, was administered in the drinking water to Sprague-Dawley rats for 2.5 mo, prior to and throughout gestation (Abdel-Rahman 1982). Concentrations were 0, 1, 10, or 100 mg/L. Rats were sacrificed on day 20 of gestation and fetuses were examined for bone and soft-tissue defects. No increase in resorptions were found in any treatment group. A significant increase in skeletal anomalies at the 100 mg/L concentration (incompletely ossified or missing sternebrae and rudimentary ribs) was interpreted as a nonspecific retardation in growth. The increase in skeletal defects was not significant in the 10 mg/L and 100 mg/L groups compared with the controls. Total defects—skeletal and soft tissue—were increased significantly over the control group in the 100 mg/L group. Maternal toxicity—body weight, food consumption, clinical signs—was not described.

3.4. Genotoxicity

No data on inhalation exposures were located in the available literature. Genotoxicity studies were conducted via oral dosing of groups of 10 male $B6C3F_1$ mice with 1 mL of test solution containing OCl^- or HOCl at concentrations of 40, 100, or 200 mg/L for 5 d (Meier et al. 1985). Chlorine was not mutagenic in the bone marrow micronucleus and cytogenetic assays. Sodium hypochlorite produced chromosomal aberrations in several mammalian cell tests (NTP 1992).

3.5. Chronic Toxicity and Carcinogenicity

Groups of male and female F-344 rats were exposed by inhalation to chlorine concentrations at 0, 0.4, 1.0, or 2.5 ppm for 2 y (CIIT 1993; Wolf et al. 1995). Histologic examinations of the nose and major organs revealed no increase in the incidence of neoplasia over that of control groups. F-344 rats and $B6C3F_1$ mice of both genders administered chlorinated or chloraminated drinking water for 2 y showed no increased incidences of neoplasms (NTP 1992).

3.6. Summary

Few animal studies addressed no- or mild-effect levels at exposure times of 10 min to 8 h. No gross or microscopic lung changes occurred in rabbits following a 30-min exposure at 50 ppm (Barrow and Smith 1975). The highest 30-min values resulting in no deaths (LC_0) for the rat and rabbit were 547 ppm (Zwart and Woutersen 1988) and 200 ppm (Barrow and Smith 1975), respectively. The 60-min concentrations resulting in no deaths in the rat and mouse were 322 (Zwart and Woutersen 1988) and 150 ppm (O'Neil 1991), respectively. No deaths, but moderate to severe lesions of the respiratory tract and peribronchiolitis, occurred in rats following a 6-h exposure at 9.1 ppm (Jiang et al. 1983).

Thirty-minute LC_{50} values ranged from 137 ppm in the mouse (Back et al. 1972) to 700 ppm in the rat (Zwart and Woutersen 1988). The 60-min LC_{50} and LC_{01} values for the rat were 455 ppm and 288 ppm (Zwart and Woutersen 1988).

Chlorine administered in the drinking water or by gavage to rats or mice did not cause reproductive or developmental problems (Druckrey 1968; Abdel-Rahman 1982; Meier et al. 1985; Carlton et al. 1986). A 2-y inhalation study with rats showed no evidence of carcinogenicity (CIIT 1993; Wolf et al. 1995). Mutagenicity tests were generally negative (Meier et al. 1985).

4. SPECIAL CONSIDERATIONS

4.1. Metabolism and Disposition

Pharmacokinetic data following acute exposures were not available. Metabolic and kinetic considerations are not relevant regarding the determination of AEGL values because animals die of acute respiratory failure. Chlorine gas reacts at the site of contact, and very little of the chemical is absorbed into the bloodstream (Eaton and Klaassen 1996).

4.2. Mechanism of Toxicity

Although of moderate solubility, chlorine is categorized as a Category I gas because it is so rapidly irreversibly reactive in the surface liquid and tissue of the respiratory tract (EPA 1994). Studies with repeated exposures of laboratory animals indicate that at moderate to high concentrations, chlorine is not effectively scrubbed in the upper respiratory tract and is therefore capable of exerting its effects over the entire respiratory tract (Barrow et al. 1979); however, at low concentrations ($\leq$2.5 ppm for up to 2 y), chlorine is effectively scrubbed in the anterior nasal passages as indicated by the absence of lesions in the lower respiratory tract of rats, mice (Wolf et al. 1995), and monkeys (Klonne et al. 1987).

Chlorine gas combines with tissue water to form hydrochloric and hypochlorous acids (HCl and HClO); the latter spontaneously breaks down into HCl and free O$^{\cdot}$, which combines with water, releasing oxygen radicals (O^-). The oxygen radical produces major tissue damage, which is enhanced by the presence of HCl (Perry et al. 1994; Wolf et al. 1995).

The response to inhalation of chlorine can range from sensory irritation and reflex bronchoconstriction to death, the latter due to pulmonary edema. The sensory irritation response to chlorine is due to stimulation of the trigeminal nerve endings in the respiratory mucosa, which results in a decrease in respiratory rate (Alarie 1981). Reflex bronchoconstriction is a local reaction in which cholinergic-like agents bind to respiratory tract cell surface receptors and trigger an increase in the intracellular concentration of cyclic guanosine monophosphate. That facilitates contraction of the smooth muscles that surround the trachea and bronchi, causing a decrease in airway diameter and a corresponding increase in resistance to airflow that may result in wheezing, coughing, a sensation of chest tightness, and dyspnea (Witschi and Last 1996). Death can occur from lack of oxygen during an asthmatic attack or if chlorine reaches the lungs and causes pulmonary edema; delayed deaths that occur starting 3 d after exposure might be due to bronchial infection (Underhill 1920; Withers and Lees 1985a; Bitron and Aharonson 1978), which may be treatable in humans.

4.3. Structure-Activity Relationships

The combined human and animal data on chlorine are sufficient for derivation of inhalation exposure guidelines and the use of structure-activity comparisons is not necessary. Like hydrogen chloride (HCl) and fluorine (F_2), Cl_2 is an irritant to the eyes, skin, and respiratory tract. When compared with mortality data for HCl and F_2 (Wohlslagel et al. 1976; ATSDR 1993; Perry et al. 1994; NAC 1996), chlorine is more toxic than HCl but slightly less toxic than F_2 to laboratory rodents.

4.4. Concentration-Exposure Duration Relationship

When considering low concentrations of chlorine that do not result in tissue damage and the response of atopic and asthmatic individuals, time-scaling might not be relevant. Several studies, such as those of Shusterman et al. (1998) and D'Alessandro et al. (1996), were conducted for short periods of time—15 min and 1 h, respectively—because responses were expected within those time periods. Furthermore, in the Rotman et al. (1983) study, pulmonary function parameters of the tested individuals, including those for the atopic individual, did not change between the 4- and 8-h measurements, indicating a sustained effect during exposure.

Time-scaling is relevant at the AEGL-2 and AEGL-3 levels where tissue damage is involved. When data are lacking for desired exposure times, scaling across time might be based on the relationship between concentration and exposure duration when a common end point is used. The relationship between concentration and time is described by $C^n \times t = k$, in which the exponent n may be different from 1. ten Berge and Vis van Heemst (1983) analyzed the data of Anglen (1981) on irritation response in humans. Irritation consisted of itching or burning of the eyes, nose, or throat and scores ranged from 0 (no sensation) to 5 (unbearable), with a score of 3 representing a nuisance level or irritation. Regression analysis of the percent of subjects reporting a nuisance irritation response to concentrations at 1 ppm and 2 ppm over exposure durations of 30 min and 120 min resulted in an n value of 1.9.

Several sets of mortality data from animal studies were available for calculating the relationship between concentration and exposure time. Using the probit analysis method of ten Berge et al. (1986) and/or regression analysis, the data sets and their values for n are Bitron and Aharonson (1978), $n = 3.5$; Zwart and Woutersen (1988), $n = 1.0$; and Weedon et al. (1940), $n = 1.1$. The probit analysis method applied to the data for 11 irritant gases from ten Berge et al. (1986) also results in a range of values for n of 1.0-3.5. In the Bitron and Aharonson (1978) study, the mice were restrained (which usually results in lower lethality values than for unrestrained animals) and deaths were delayed, occurring during the second week of the observation period rather than during and immediately following exposure. Because of the questionable methodology and the delayed deaths (possibly due to bacterial infection) in the Bitron and Aharonson (1978) study, that study will not be considered for calculating a time-scaling relationship for chlorine. Respiratory irritation is an initial step in the progression of irritation that leads to pulmonary edema and death. Based on the evidence for a similar mechanism of action for irritation and death, an n value of 2 will be used for time-scaling for chlorine. An n value of 2 for time-scaling the threshold for lethality for chlorine is supported by data for another halogen. Using the end point of lethality, the concentration-exposure duration relationship for fluorine for several mammalian species was $C^n \times t = k$, where n was approximately slightly less than 2 (NAC 1996). Based on relative toxicity (Section 4.3), the n value for chlorine would be higher than that of fluorine when extrapolating from longer to shorter exposure durations.

4.5. Other Relevant Information

4.5.1. Susceptible Populations

Chlorine is highly irritating and corrosive to the tissues of the respiratory tract. At low concentrations, the direct action of chlorine on the respiratory tract is not expected to vary greatly among most healthy individuals, including infants, children, and the elderly. For example, at the low concentration of 0.5 ppm, neither healthy nor atopic subjects exhibited clinically significant changes (decreases of ≥ 10%) in peak air flow (Shusterman et al. 1998). At 0.4 ppm, there were no statistically significant changes in several respiratory parameters in either healthy or asthmatic subjects (D'Alessandro et al. 1996). In the Rotman et al. (1983) study where numerous pulmonary parameters were measured, healthy subjects responded in a similar manner at 1 ppm.

Data from the Rotman et al. (1983) and D'Alessandro et al. (1996) studies on chlorine exposures and individuals with airway hyper-reactivity or asthma indicate that, compared with the general population, the respiratory tracts of those individuals may be very reactive to the presence of chlorine, as reported in Section 2.2.1. Responsiveness to inhaled agents varies among individuals with airway hyper-reactivity and asthma. In children, asthma may be defined as mild, moderate, or severe depending on the response to an inhaled agonist (Larsen 1992). With mild asthma, symptoms are infrequent and brief, and treatment is with inhaled β-agonists as needed. Moderate asthma is defined by symptoms occurring twice weekly, and treatment is with cromolyn sodium or slow-release theophylline. Severe asthmatic patients have daily symptoms, and daily treatment with oral or inhaled steroids is required. One individual in the D'Alessandro et al. (1996) study with airway hyper-reactivity had a clinical history of asthma and "was being treated regularly with inhaled or systemic corticosteroids."

There is a concern that children with airway hyper-reactivity and asthma may be more sensitive to inhaled irritants and allergens than adult asthmatic subjects. Avital et al. (1991) studied the response to methacholine challenge in 182 asthmatic children (132 males and 50 females) of various ages. All therapy except corticosteroid therapy and slow-release theophylline was discontinued prior to the study. The children were divided into three age groups—1-6 y, 7-11 y, and 12-17 y—and into three clinical groups according to their minimal therapeutic requirements—mild, moderate, and severe asthma. In older children, responsiveness was measured by the methacholine provocation concentration that produced a 20% fall in FEV_1. In younger children, responsiveness was determined by wheezing, persistent cough, or tachypnea following a methacholine challenge. Bronchial reactivity correlated inversely with the severity of bronchial asthma according to minimal drug requirements and was similar over the age range for each severity group. That is, in each severity category (mild, moderate, or severe asthma), the mean concentration of methacholine that evoked the designated response was the same regardless of the age group. Avital et al. (1991) compared their results with the responses of asthmatic adults challenged with histamine from a study by Cockcroft et al. (1977). Mean ages of the adults were between 30 y and 40 y. Although the classes of asthmatic severity in Cockcroft et al. (1977) were only broadly comparable to those in Avital et al. (1991), there was a "striking similarity" in the results. That is, the concentration at which adults in each class of severity responded to histamine was similar to the methacholine concentration at which the children reacted in the respective severity classes.

Adults were tested in the D'Alessandro et al. (1996) study, but the ages of the individuals in the lower range of age (18 y) were close to those of the older children in the Avital et al. (1991) study. The range of provocative concentrations of methacholine in the D'Alessandro et al. (1996) study overlapped the range of the severe asthmatic subjects in the Avital et al. (1991) study, indicating that at least one adult in the former study was as responsive to methacholine as the children with severe asthma. Furthermore, during the exposure to chlorine at 1 ppm, the atopic individual in the Rotman et al. (1983) study, who was not on medication, responded to a greater degree, as measured by a fall in FEV_1, than any of the subjects with airway reactivity or asthma in the D'Alessandro et al. (1996) study. Following a 4-h exposure to chlorine at 1 ppm, the FEV_1 of the atopic individual was 45% of the pre-exposure value (Rotman et al. 1983), whereas following the 1-h exposure at 1 ppm, the FEV_1 of the most sensitive individual with airway hyper-reactivity or asthma was 61% of the pre-exposure value (D'Alessandro et al. 1996). The final FEV_1 for both subjects was the same, 1,900 mL.

Another consideration when evaluating the response of individuals with airway hyper-reactivity and asthma to irritants and allergens is the time of response following challenge. Individuals in the D'Alessandro et al. (1996) study were exposed to chlorine for only 1 h; the response of the atopic individual in the Rotman et al. (1983) study may have occurred early, although wheezing did not occur until after 4 h of exposure. According to the literature, asthmatic reactions may be immediate (within minutes of exposure) or delayed (hours after exposure) (Larsen 1992). In the Avital et al. (1991) study, response to provocative concentrations of methacholine occurred within 2 min. In both studies with chlorine (Rotman et al. 1983; D'Alessandro et al. 1996), individuals had responded by the time they were tested for pulmonary function changes (after 1 h and after 4 h of exposure), and there were no reported delayed or greater effects within the 24-h post-exposure period. Although not conclusive, the data appear to indicate that the response to chlorine is concentration-dependent rather than time-dependent. Thus, a 1-h exposure to chlorine is sufficient to elicit a response in susceptible individuals; if a response is not present at 1 h, it is unlikely to occur with continued exposure.

4.5.2. Reactive Airways Dysfunction Syndrome

In humans, the reported long-term effects of accidental exposures at high concentrations of chlorine (evidenced by the presence of a yellow-green cloud) are conflicting, some authors noting residual pulmonary abnormalities (Kowitz et al. 1967; Alberts and do Pico 1996), and others either reporting no significant permanent damage (Weill et al. 1969; Kaufman and Burkons 1971; Jones et al. 1986) or reporting that the presence of permanent damage was questionable (Charan et al. 1985). Several case reports described respiratory hyper-responsiveness following acute exposures to chlorine at high concentrations. This syndrome, called reactive airways disease or reactive airways dysfunction syndrome (RADS), is initiated by one or several exposures to high concentrations of an irritating gas. Case studies were reviewed by Alberts and do Pico (1996) and Lemiere et al. (1996). In several of the studies, a clear interpretation of the results was complicated by the lower values in pulmonary function tests of smokers.

4.5.3. Gender and Species Variability

Several studies indicated gender differences in responses to chlorine exposure, females rats (Barrow et al. 1979; Wolf et al. 1995) and monkeys (Klonne et al. 1987) developing lesions at lower concentrations than males. However, male mice were slightly more sensitive to chlorine than female mice (Wolf et al. 1995).

The data also allow an examination of interspecies differences. In those studies in which investigators tested the lethality of chlorine to two species (rat and mouse), the LC_{50} values were within a factor of approximately 2 of each other (Weedon et al. 1940; Back et al. 1972; Zwart and Woutersen 1988).

4.5.4. Tolerance to Repeated Exposures

Following repeated exposures, rats developed sensory irritation tolerance to chlorine, as indicated by an increase in the RD_{50} following pretreatment (Barrow and Steinhagen 1982). The RD_{50} was 25 ppm in non-pretreated rats, whereas RD_{50} values increased to 90, 71, and 454 ppm when rats were pretreated at 1, 5, or 10 ppm, respectively, for 6 h/d, 5 d/wk for 2 wk and exposed 16-24 h following the last day of pretreatment.

5. DATA ANALYSIS FOR AEGL-1

5.1. Summary of Human Data Relevant to AEGL-1

NIOSH (1976) discussed available epidemiology studies conducted prior to 1976. In those studies, and in a more recent study (Ferris et al. 1979), work room concentrations averaged <1 ppm, and no effects could be clearly documented. However, those studies generally included a "healthy worker" population.

The studies by Anglen (1981) and Rotman et al. (1983) indicate that there are no significant sensory irritation and no serious pulmonary function changes associated with 4- or 8-h exposures in healthy human subjects of both genders at 0.5 ppm or 1.0 ppm. However, an atopic individual, whose baseline pulmonary parameters were outside of the normal range, as defined by DuBois et al. (1971), experienced changes in several pulmonary function parameters after exposure to chlorine at 0.5 ppm (Rotman et al. 1983). The greatest change for that individual was in R_{aw}, which increased 40% over the pre-exposure value after 4 h of exposure at 0.5 ppm and increased 33% over the pre-exposure value after 8 h of exposure at 0.5 ppm. Changes in R_{aw} in the healthy subjects were 5% and 15% for the respective time periods. Transient changes in pulmonary function (specifically R_{aw}) were seen at 1.0 ppm for 4 h in the Rotman et al. study (1983) and at 2 ppm for 4 h in the Anglen (1981) study; however, sensory irritation reached nuisance levels at the latter concentration-exposure time combination. In healthy subjects, no lung function measurements were changed at 2.0 ppm for 2 h (Anglen 1981). The FEV_1, a particularly reproducible and sensitive measure of obstructive or restrictive flows in the lung (Witschi and Last 1996), changed by less than 10% following exposure at 1 ppm for 4 h. The atopic individual did not tolerate the 1-ppm exposure during a second 4-h exposure period because of serious respiratory effects.

In the study by D'Alessandro et al. (1996), subjects with a clinical history of asthma were tested. In the study by Shusterman et al. (1998), subjects with seasonal allergic rhinitis were tested. A concentration at 0.4 ppm for 1 h elicited no statistically significant response in airflow or resistance in asthmatic subjects (D'Alessandro et al. 1996). In the same study, a concentration at 1.0 ppm for 1 h elicited significant changes in several pulmonary function parameters for both normal subjects and subjects with asthma, but the mean changes were considered modest by the study authors. However, the R_{aw} of one subject with asthma more than tripled during the exposure at 1.0 ppm.

5.2. Summary of Animal Data Relevant to AEGL-1

All short-term animal studies were conducted at concentrations that produced effects greater than those defined by the AEGL-1.

5.3. Derivation of AEGL-1

The studies by Anglen (1981), Rotman et al. (1983), D'Alessandro et al. (1996), and Shusterman et al. (1998) are all relevant to the development of AEGL-1 values. Those studies addressed sensory irritation as well as symptomatic and asymptomatic changes in several pulmonary function parameters. In addition, the studies used a diverse population, including healthy, atopic, and asthmatic subjects. Exercise was incorporated into the protocol of the Rotman et al. (1983) study, simulating conditions of stress. The studies indicate that 0.5 ppm, for 15 min (Shusterman et al. 1998) or for an interrupted 8 h with incorporation of exercise into the protocol (Rotman et al. 1983), is the highest NOAEL consistent with the definition of an AEGL-1 (i.e., a NOAEL for notable discomfort and irritation accompanied by non-disabling, transient, asymptomatic effects). The next highest concentration tested, 1.0 ppm for more than 4 h, resulted in effects, such as shortness of breath and wheezing, greater than those defined by an AEGL-1. The NOAEL of 0.5 ppm was identified from the study of Rotman et al.

subjects in the study of D'Alessandro et al. (1996). That NOAEL supports the choice of 0.5 ppm in the study of Rotman et al. (1983). In fact, the single exercising atopic subject proved to be more sensitive at 0.5 ppm, as indicated by asymptomatic changes in pulmonary function parameters, than were the non-exercising asthmatics exposed at 0.4 ppm.

Because subjects identified as most susceptible to the irritant effects of chlorine were tested (atopic and asthmatic individuals), an intraspecies uncertainty factor (UF) of 1 was applied. The intraspecies UF of 1 is also supported by the fact that bronchoconstriction is induced in pediatric and adult asthmatic subjects at similar levels of challenge (Avital et al. 1991). Therefore, an additional UF to protect pediatric asthmatic subjects is not necessary. Time-scaling was not applied to the AEGL-1 for several reasons. The study conducted by Rotman et al. (1983) actually lasted more than 8 h (two 4-h sessions with a 1-h break between). That reduces the uncertainty usually associated with scaling from shorter to longer time periods. Because effects were not increased following an interrupted 8 h of exposure at 0.5 ppm in the susceptible individual, the 8-h AEGL-1 was also set at 0.5 ppm. The use of the same value across all exposure durations is supported by the fact that the response to the irritant effects of chlorine appears to be concentration-dependent rather than time-dependent. Calculations are presented in Appendix A; results are presented in Table 1-6. Figure 1-1 is a plot of the derived AEGL values and all of the human and animal data on chlorine.

6. DATA ANALYSIS FOR AEGL-2

6.1. Summary of Human Data Relevant to AEGL-2

The studies by Joosting and Verberk (1974), Anglen (1981), Rotman et al. (1983), and D'Alessandro et al. (1996) address sensory irritant effects in humans as well as differences in pulmonary function tests during exposures at 1.0 ppm for up to 8 h and 2.0 ppm and 4.0 ppm for 4 h. As previously noted, Rotman et al. (1983) used an exercising atopic individual, and D'Alessandro et al. (1996) used five subjects with nonspecific airway hyper-reactivity, three of which had clinical histories of asthma. For healthy individuals, the concentrations tested by Rotman et al. (1983) and D'Alessandro et al. (1996) resulted in effects below the definition of the AEGL-2. However, in one atopic subject, the 1-ppm concentration for

TABLE 1-6 AEGL-1 Values for Chlorine (ppm [mg/m^3])

10 min	30 min	1 h	4 h	8 h
0.5 (1.5)	0.5 (1.5)	0.5 (1.5)	0.5 (1.5)	0.5 (1.5)

more than 4 h resulted in shortness of breath and wheezing as well as serious pulmonary function changes (Rotman et al. 1983). In the D'Alessandro et al. (1996) study, there was a significant fall in FEV_1 and airway resistance immediately after exposure at 1.0 ppm in both healthy and asthmatic subjects, but the fall among asthmatic subjects was greater. Two of the subjects with airway hyper-reactivity experienced undefined respiratory symptoms following the exposure. Those subjects were not specifically identified as asthmatics. The 2-ppm concentration tested by Anglen (1981) reached nuisance levels by 4 h and was accompanied by transient changes in pulmonary functions. Joosting and Verberk (1974) did not find differences in pulmonary function after exposure at 2 ppm for 2 h; concentrations at 4 ppm for 2 h were irritating, but pulmonary function measurements were not made. Subjective sensory irritation of the throat reached an average level of "nuisance" (distinctly perceptible to offensive) at 4 ppm for 2 h; irritation of the nose and the desire to cough averaged "distinctly perceptible" (range, up to nuisance). No sensory response was reported as unbearable during these exposures. Although airflow resistance, the major pulmonary effect found in the Rotman et al. (1983) study, was not measured by Joosting and Verberk (1974), Anglen (1981) reported that the changes in other pulmonary parameters at 2 ppm for 4 or 8 h were completely reversible by the next day.

6.2. Summary of Animal Data Relevant to AEGL-2

Alarie (1981), the author of the ASTM (1981) RD_{50} test, considered the RD_{50} for male Swiss-Webster mice to be intolerable to humans but nonlethal over a period of hours. The RD_{50} of male Swiss-Webster mice was 9.3 ppm, as reported by Barrow et al. (1977). Lesions were present in the nasal passages, and some changes were present in the lower respiratory tract of rats exposed at 9.1 ppm for 6 h (Jiang et al. 1983). Mice "tolerated" that concentration 6 h/d for 5 d without deaths (Buckley et al. 1984). No gross or microscopic lung changes occurred in rabbits following a 30-min exposure at 50 ppm (Barrow and Smith 1975).

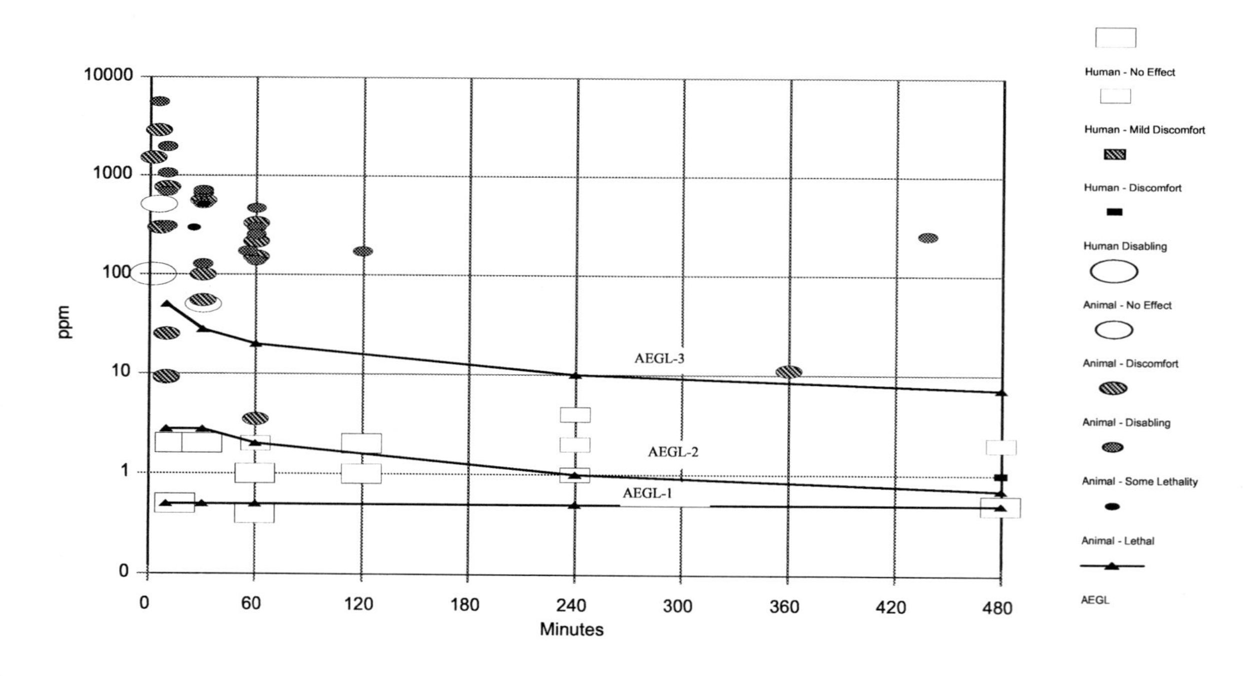

FIGURE 1-1 Toxicity data and AEGL values for chlorine. Toxicity data include both human and animal studies.

TABLE 1-7 AEGL-2 Values for Chlorine (ppm [mg/m^3])

10 min	30 min	1 h	4 h	8 h
2.8 (8.1)	2.8 (8.1)	2.0 (5.8)	1.0 (2.9)	0.71 (2.0)

6.3. Derivation of AEGL-2

Because human data are available, they should be used to calculate the AEGL-2. In the Rotman et al. (1983) study, the exposure of a susceptible subject to chlorine at 1 ppm for 4 h did not produce respiratory symptoms, but an exposure for more than 4 h resulted in serious asthma-like symptoms and serious pulmonary function changes. The pulmonary function changes were reversible, and the ability to escape was not impaired, but an asthma-like attack that occurred during the longer exposure must be considered a serious health effect. Therefore, the first 4-h exposure, which was without symptoms, can be considered a NOAEL for the symptoms that define AEGL-2. In the D'Alessandro et al. study (1996), two subjects with airway hyper-reactivity and three subjects with asthma also experienced significant changes in several pulmonary function parameters following a 1-h exposure at 1.0 ppm. No respiratory symptoms were experienced during the exposure, although undefined symptoms were experienced later. Because individuals representative of the susceptible population were tested and their reactions met the definition of the AEGL-2, no UF for differences in human sensitivity was applied. Time-scaling was applied to the AEGL-2 values because the 1.0-ppm concentration meets the definition of nuisance irritation, as described by ten Berge and Vis van Heemst (1983), and evacuation or sheltering should take place at the high short-term exposure concentrations that might cause an asthma attack. The 4-h 1-ppm concentration was scaled to the other time periods using the $C^2 \times t = k$ relationship, derived by ten Berge and Vis van Heemst (1983). The 10-min value was set equal to the 30-min value so that the highest exposure of 4.0 ppm in the controlled human studies was not exceeded. Results are presented in Table 1-7 (above), and the calculations are presented in Appendix A.

The 1-y study in which exposure of Rhesus monkeys to a concentration of Cl_2 at 2.3 ppm resulted in mild histopathologic changes in the nasal passages and trachea (Klonne et al. 1987) supports the safety of this short-term exposure for humans. Regarding size, anatomy, and tissue distribution, the respiratory tract of the monkey is an appropriate model for humans.

7. DATA ANALYSIS FOR AEGL-3

7.1. Summary of Human Data Relevant to AEGL-3

Although exposure concentrations as high as 30 ppm (actual exposure concentration and duration not known) and 66 ppm for unspecified amounts of time have been reported by Abhyankar et al. (1989) and Shroff et al. (1988), respectively, measurement methods and/or exposure times were not well-documented. No deaths were reported at those exposures. No reliable data on concentrations that caused death in humans were located.

7.2. Summary of Animal Data Relevant to AEGL-3

The most recent available data for relevant exposure times show that 1-h concentrations that resulted in no deaths range from 150 ppm in the mouse (O'Neil 1991) to 322 ppm in the rat (Zwart and Woutersen 1988). On the basis of the rat data, Zwart and Woutersen (1988) calculated an LC_{01} of 288 ppm. Thirty-minute concentrations that resulted in no deaths were 547 ppm in the rat (Zwart and Woutersen 1988) and 200 ppm in the rabbit (Barrow and Smith 1975). The mouse data of Zwart and Woutersen (1988) were insufficient for calculating an LC_{01}. In addition, lethality measurements in many of the mouse studies were complicated by delayed deaths (Bitron and Aharonson 1978; O'Neil 1991), which were attributed to pneumonia. Mice might not be an appropriate model for extrapolation to humans, as stated by ten Berge et al. (1986) and the NRC (1991) for HCl. The NRC (1991) states that when considering lethal concentrations of respiratory irritants (such as HCl), the mouse "may not be an appropriate model for extrapolation to humans," because "mice appear to be much more susceptible to the lethal effects of HCl than other rodents or baboons. To some extent, this increased susceptibility may be due to less effective scrubbing of HCl in the upper respiratory tract."

7.3. Derivation of AEGL-3

Because the experimental data in mice appeared to provide an overly conservative estimate of lethality that was not consistent with the overall preponderance of the data, a value less than the concentration that resulted in no deaths in rats but greater than the value that resulted in no deaths in mice was chosen as the basis for the AEGL-3 values. The 200-ppm value is below the 1-h highest nonlethal concentrations (213 ppm and 322 ppm) and the LC_{01} (288 ppm) in two well-conducted studies with rats (MacEwen and Vernot 1972; Zwart and Woutersen 1988) and above the 1-h highest nonlethal concentration in mice, 150 ppm (O'Neil 1991). The 200-ppm concentration is an LC_{20} for the mouse. A UF of 3 was used to extrapolate from rats to humans (the data show that interspecies differences were within a factor of approximately 2 for lethality). In addition, chlorine is a contact-site, direct-acting toxicant; there is no metabolic or pharmacokinetic component to chlorine-induced effects, and there is likely to be little difference between species in the response of biologic tissues to chlorine exposure. Also, for intraspecies differences, corrosive gases acting at the point of contact would predict low variability in a population; thus a UF of 3 is applied to protect susceptible individuals. The relative sensitivity of asthmatic individuals when considering lethality is unknown, but the data from the Rotman et al. (1983) and D'Alessandro et al. (1996) studies show that doubling a no-effect concentration for irritation and bronchial constriction resulted in potentially serious effects. The data were divided by a combined UF of 10 and were scaled to the 30-min and 4- and 8-h exposure durations using the $C^2 \times t = k$ relationship, which was based on a nuisance level of human irritation (ten Berge and Vis van Heemst 1983). Application of a larger intraspecies UF, such as 10 (for a total of 30), would result in an 8-h AEGL-3 value of 2.3 ppm. It is unlikely that asthmatic subjects would suffer life-threatening symptoms at 2.3 ppm. The values appear in Table 1-8 and the calculations are presented in Appendix A.

It should be noted that the 1-h AEGL-3 of 20 ppm is below the AEGL-3 derived from the 1-h rat LC_{01} (the threshold for lethality) of 28.8 ppm (288/10).

8. SUMMARY OF AEGLs

8.1. AEGL Values and Toxicity End Points

In summary, the AEGL values for various levels of effects and various exposure periods were derived using the following methods. The AEGL-1 was based on a study with human volunteers, including a susceptible individual, in which a concentration of chlorine at 0.5 ppm for 4 h produced no sensory irritation and resulted in only mild transient effects on pulmonary parameters in the healthy individuals. Pulmonary changes in the susceptible

TABLE 1-8 AEGL-3 Values for Chlorine (ppm [mg/m^3])

10 min	30 min	1 h	4 h	8 h
50 (145)	28 (81)	20 (58)	10 (29)	7.1 (21)

individual were greater than those in healthy subjects, but did not result in symptoms above the definition of the AEGL-1. Because both genders were tested in one of the studies, a susceptible individual was observed in the other study, and subjects were undergoing light exercise (making them more vulnerable to sensory irritation), no UF to account for differences in human sensitivity was applied. The 0.5-ppm no-effect concentration for susceptible individuals is supported by a 1-h 0.4-ppm no-effect concentration for individuals with airway hyper-reactivity or asthma. The 0.5-ppm exposure was considered a threshold for more severe effects, regardless of exposure duration.

The AEGL-2 values were based on the same studies used to derive the AEGL-1s. In those studies healthy human volunteers experienced transient changes in pulmonary function measurements and a susceptible individual experienced an asthma-like attack (shortness of breath and wheezing) following a more than 4-h exposure to chlorine at 1 ppm. The susceptible individual remained in the exposure chamber for the full 4 h before the symptoms occurred. Because both genders were tested, subjects were undergoing light exercise (making them more vulnerable to sensory irritation), and a susceptible individual was tested, no UF was applied to account for differences in human sensitivity. Similar effects and symptoms in individuals with airway hyper-reactivity or asthma exposed at 1.0 ppm for 1 h supports the application of no intraspecies UF for the 4-h concentration. The 4-h 1-ppm concentration was scaled to the other time periods using the $C^2 \times t = k$ relationship.

In the absence of human data, the AEGL-3 values were based on a concentration lower than the highest value that resulted in no deaths for the rat but equal to the LC_{20} for the mouse. The data resulting from exposure of mice to chlorine were either incomplete or complicated by delayed deaths. The 1-h concentration of 200 ppm was divided by a UF of 3 to extrapolate from animals to humans (interspecies values for the same end point differed by a factor of approximately 2 within each of several studies) and by a UF of 3 to account for differences in human sensitivity (the toxic effect is the result of a chemical reaction with biologic tissue of the respiratory tract, which is unlikely to differ among individuals). The AEGL-3 values for the other exposure times were calculated based on the $C^2 \times t = k$ relationship.

The three AEGL levels for the four exposure times are listed in Table 1-9.

8.2. Extant Standards and Guidelines for Chlorine

Standards and guidance levels for workplace and community exposures are listed in Table 1-10. The 8-h AEGL-1 and the ACGIH TLV-TWA values are the same, and, although not specifically stated, the TLV-TWA is based on the absence of irritation during the 8-h exposure at 0.5 ppm in the Anglen (1981) and Rotman et al. (1983) studies. Those and another, more recent study are also the basis for the AEGL-1 and AEGL-2 values. The AEGL-1 is specifically protective of asthmatic subjects.

The NIOSH IDLH is based on acute inhalation toxicity data in humans, Freitag (1941, as cited in NIOSH 1994a), and several secondary sources, such as ILO (1998). Exposure at 30 ppm is stated to cause intense coughing fits, and exposure at 40-60 ppm for 30-60 min might cause serious damage (ILO 1998). The IDLH is higher than the AEGL-2, but less than the AEGL-3. The AEGL-2 is lower because it is specifically protective of asthmatic patients.

The ERPG-1 is based on an 8-h human exposure at 1 ppm that produced slight transient pulmonary effects in healthy subjects (Anglen 1981; Rotman et al. 1983). The 1-h AEGL-1 is lower than the ERPG-1 because

TABLE 1-9 Summary and Relationship of AEGL Values

	Exposure Duration				
Classification	10 min	30 min	1 h	4 h	8 h
AEGL-1 (Nondisabling)	0.5 ppm (1.5 mg/m^3)	0.5 ppm (1.5 mg/m^3)	0.5 ppm (1.5 mg/m^3)	0.5 ppm (1.5 mg/m^3)	0.5 ppm (1.5 mg/m^3)
AEGL-2 (Disabling)	2.8 ppm (8.1 mg/m^3)	2.8 ppm (8.1 mg/m^3)	2.0 ppm (5.8 mg/m^3)	1.0 ppm (2.9 mg/m^3)	0.71 ppm (2.0 mg/m^3)
AEGL-3 (Lethal)	50 ppm (145 mg/m^3)	28 ppm (81 mg/m^3)	20 ppm (58 mg/m^3)	10 ppm (29 mg/m^3)	7.1 ppm (21 mg/m^3)

Abbreviations: mg/m^3, milligrams per cubic meter; ppm, parts per million.

TABLE 1-10 Extant Standards and Guidelines for Chlorine (ppm)

	Exposure Duration				
Guideline	10 min	30 min	1 h	4 h	8 h
AEGL-1	0.5	0.5	0.5	0.5	0.5
AEGL-2	2.8	2.8	2.0	1.0	0.71
AEGL-3	50	28	20	10	7.1
ERPG-1 (AIHA)[a]			1		
ERPG-2 (AIHA)			3		
ERPG-3 (AIHA)			20		
EEGL (NRC)[b]					3
PEL-Ceiling (OSHA)[c]					1
IDLH (NIOSH)[d]		10			
REL-Ceiling (NIOSH)[e]					0.5
TLV-TWA (ACGIH)[f]					0.5
TLV-STEL (ACGIH)[g]					1.0
MAK (Germany)[h]					0.5
MAK - Peak Limit (Germany)[i]					1.0
MAC - Ceiling (The Netherlands)[j]					1.0

[a]ERPG (emergency response planning guidelines) of the American Industrial Hygiene Association) (AIHA 2001). ERPG-1 is the maximum airborne concentration below which it is believed nearly all individuals could be exposed for up to 1 h without experiencing symptoms other than mild, transient adverse health effects or without perceiving a clearly defined objectionable odor. The ERPG-1 for chlorine is based on the transient effects on pulmonary function parameters observed during exposure at 1.0 ppm in the studies of Anglen (1981) and Rotman et al. (1983). The ERPG-2 is the maximum airborne concentration below which it is believed nearly all individuals could be exposed for up to 1 h without experiencing or developing irreversible or other serious health effects or symptoms that could impair an individual's ability to take protection action. The ERPG-2 for chlorine is based on the

human data and the subchronic and chronic animal studies. The ERPG-3 is the maximum airborne concentration below which it is believed nearly all individuals could be exposed for up to 1 h without experiencing or developing life-threatening health effects. The ERPG-3 for chlorine is based on scientific judgment; the studies of Schlagbauer and Henschler (1967, as cited in AIHA 1988) and Withers and Lees (1985) are mentioned.

[b]EEGL (emergency exposure guidance level) (NRC 1985). The EEGL is the concentration of a contaminant that can cause discomfort or other evidence of irritation or intoxication in or around the workplace, but avoids death, other severe acute effects, and long-term or chronic injury. EEGLs were developed for healthy military personnel. The EEL (emergency exposure limit) for chlorine was set in 1966 and 1971 on the basis of nose and eye irritation. After review of the Rotman et al. (1983) data, the 8-h EEL was kept at 3 ppm, but the 24-h EEGL was lowered to 0.5 ppm.

[c]OSHA PEL–ceiling (permissible exposure limit–ceiling of the Occupational Safety and Health Administration) (OSHA 1997). The PEL–ceiling should not be exceeded at any dime during a work day (instantaneous monitoring or 15-min TWA).

[d]IDLH (immediately dangerous to life and health standard of the National Institute of Occupational Safety and Health) (NIOSH 1994). The IDLH represents the maximum concentration from which one could escape within 30 min without any escape-impairing symptoms or irreversible health effects. The IDLH for chlorine is based on acute inhalation toxicity data in humans, specifically Freitag (1941, as cited in NIOSH 1994a) and several secondary sources.

[e]NIOSH REL-ceiling (recommended exposure limits–ceiling) (NIOSH 1997). The REL-ceiling should not be exceeded at any time during a work day.

[f]ACGIH TLV–TWA (Threshold Limit Value–time-weighted average of the American Conference of Governmental Industrial Hygienists) (ACGIH 2000). The time-weighted average concentration for an 8-h work day and a 40-h work week to which nearly all workers may be repeatedly exposed, day after day, without adverse effects. The TWA is based on the absence of pulmonary effects at a concentration of chlorine at 0.5 ppm (from Rotman et al. 1983).

[g]ACGIH TLV–STEL (Threshold Limit Value–short-term exposure limit) (ACGIH 2001). A 15-min TWA exposure that should not be exceeded at any time during the work day even if the 8-h TWA is within the TLV– TWA. Exposures above the TLV–TWA up to the STEL should not be longer than 15 min and should not occur more than 4 times per day. There should be at least 60 min between successive exposures in this range. The STEL is based on significant changes in pulmonary function parameters during exposure at 1.0 ppm observed in the Rotman et al. (1983) study.

[h]MAK (Maximale Argeitsplatzkonzentration [Maximum Workplace Concentration]) (Deutsche Forschungsgemeinschaft [German Research Association] 2000). The MAK is defined analogous to the ACGIH TLV–TWA.

[i]MAK Spitzenbegrenzung (peak limit) (German Research Association 2001). The peak limit constitutes the maximum average concentration to which workers can be

exposed for a period up to 30 min with no more than two exposure periods per work shift; total exposure may not exceed the 8-h MAK.
[j]MAC (Maximaal Aanvaaarde Concentratie [Maximal Accepted Concentration]) (SDU Uitgevers [under the auspices of the Ministry of Social Affairs and Employment], The Hague, The Netherlands 2000). The MAC is defined analogous to the ACGIH TLV–TWA.

it is specifically protective of the asthmatic population. The ERPG-2 is based on human data and subchronic and chronic animal data (unreferenced) and is only slightly higher than the 1-h AEGL-2. The ERPG-3 is based on a preponderance of the animal data (unreferenced) and is the same as the 1-h AEGL-3.

8.3. Data Adequacy and Research Needs

The human database is extensive (Joosting and Verberk 1974; Anglen 1981; Rotman et al.1983; D'Alessandro et al. 1996; Shusterman et al. 1998) and addresses the healthy population as well as asthmatic and atopic individuals. Both genders were tested during exercise in the Anglen (1981) study, and both sensory irritation and pulmonary function parameters were measured. Rotman et al. (1983) measured a range of pulmonary function parameters and included a susceptible individual. D'Alessandro et al. (1996) tested subjects with airway hyper-reactivity or asthma. In addition, in the study by Anglen (1981), exposure concentrations were measured by several different methods, all of which gave similar results. Exposure concentrations were similarly measured in the Rotman et al. (1983) study. Animal studies involved single and multiple species acute tests with multiple dosing regimens and indicated a clear dose-response relationship. Longer-term animal studies that can be used to support the safety of acute exposures were also available. At the higher concentrations, tissue and organ pathology indicated the same toxic mechanism across species.

The mammalian lethality database is extensive and includes four species and exposure times of 5 min to 16 h, as well as chronic studies. However, some of the studies were judged to suffer from methodology and concentration-analysis shortcomings. The mouse appeared to be more sensitive to chlorine than the rat, and in several studies, lethality may have been due to bronchopneumonia, a treatable effect in humans. When developing emergency exposure guidance levels (EEGLs) for hydrogen chloride, also an irritant chemical, the NRC did not consider the mouse an appropriate model for extrapolations to humans. Therefore, basing the AEGL-3 levels on a concentration higher than the highest nonlethal level in mice but lower than the highest nonlethal concentration in rats is considered appropriate.

9. REFERENCES

Abdel-Rahman, M.S., M.R. Berardi, and R.J. Bull. 1982. Effect of chlorine and monochloramine in drinking water on the developing rat fetus. J. Appl. Toxicol. 2:156-159.

Abhyankar, A., N. Bhambure, N.N. Kamath, S.P. Pajankar, S.T. Nabar, A. Shrenivas, A.C. Shah, and S.N. Deshmukh. 1989. Six month follow-up of fourteen victims with short-term exposure to chlorine gas. J. Soc. Occup. Med. 39:131-132.

ACGIH (American Conference of Governmental Industrial Hygienists). 2001. Documentation of the Threshold Limit Values and Biological Exposure Indices: Chlorine, 6th Ed. Cincinnati, OH: ACGIH.

ACGIH (American Conference of Governmental Industrial Hygienists). 2001. Threshold Limit Values (TLVs) for Chemical and Physical Agents and Biological Exposure Indices (BEIs). Cincinnati, OH: ACGIH.

AIHA (American Industrial Hygiene Association). 1988. Emergency Response Planning Guidelines, Chlorine. Akron, OH: AIHA.

Alarie, Y. 1980. Toxicological evaluation of airborne chemical irritants and allergens using respiratory reflex reactions. Pp. 207-231 in Proceedings, Symposium on Inhalation Toxicology and Technology. Ann Arbor, MI: Ann Arbor Science.

Alarie, Y. 1981. Dose-response analysis in animal studies: Prediction of human responses. Environ. Health Perspect. 42:9-13.

Alberts, W.M., and G.A. do Pico. 1996. Reactive airways dysfunction syndrome. Chest 109:1618-1626.

Amoore, J.E., and E. Hautala. 1983. Odor as an aid to chemical safety: Odor thresholds compared with Threshold Limit Values and volatilities for 214 industrial chemicals in air and water dilution. J. Appl. Toxicol. 3:272-290.

Anglen, D.M. 1981. Sensory response of human subjects to chlorine in air. Ph.D. Dissertation, University of Michigan.

Arlong, F., E. Berthet, and J. Viallier. 1940. Action of chronic intoxication by low concentration chlorine fumes on experimental guinea pigs. Presse Med. 48:361.

ASTM (American Society for Testing and Materials). 1991. Standard test method for estimating sensory irritancy of airborne chemicals. In Annual Book of ASTM Standards, Vol. 11.04. Philadelphia, PA: ASTM.

ATSDR (Agency for Toxic Substances and Disease Registry). 1993. Fluorides, Hydrogen Fluoride, and Fluorine (F). U.S. Department of Health and Human Services, Public Health Service, Atlanta, GA.

Avital, A., N. Noviski, E. Bar-Yishay, C. Springer, M. Levy, and S. Godfrey. 1991. Nonspecific bronchial reactivity in asthmatic children depends on severity but not on age. Am. Rev. Respir. Dis. 144:36-38.

Back, K.C., A.A. Thomas, and J.D. MacEwen. 1972. Reclassification of Materials Listed as Transportation Health Hazards. Report No. TSA-20-72-3. Aerospace Medical Research Laboratory, Wright-Patterson AFB, Dayton, OH.

Barrow, C.S., and W.H. Steinhagen. 1982. Sensory irritation tolerance development to chlorine in F-344 rats following repeated inhalation. Toxicol. Appl. Pharmacol. 65:383-9.

Barrow, C.S., Y. Alarie, J.C. Warrick, and M.F. Stock. 1977. Comparison of the sensory irritation response in mice to chlorine and hydrogen chloride. Arch. Environ. Health 32:68-76.

Barrow, C.S., R.J. Kociba, L.W. Rampy, D.G. Keyes, and R.R. Albee. 1979. An inhalation toxicity study of chlorine in Fischer-344 rats following 30 days of exposure. Toxicol. Appl. Pharmacol. 49:77-88.

Barrow, R.E., and R.G. Smith. 1975. Chlorine induced pulmonary function changes in rabbits. Am. Ind. Hyg. Assoc. J. 36:398-403.

Bell, D.P., and P.C. Elmes. 1965. The effects of chlorine gas on the lungs of rats without spontaneous pulmonary disease. J. Pathol. Bacteriol. 89:307-317.

Bitron, M.D., and E.F. Aharonson. 1978. Delayed mortality of mice following inhalation of acute doses of formaldehyde, sulfur dioxide, chlorine and bromine. Am. Ind. Hyg. Assoc. J. 39:129-138.

Buckley, L.A., X.Z. Jiang, R.A. James, K.T. Morgan, and C.S. Barrow. 1984. Respiratory tract lesions induced by sensory irritants ant the RD_{50} concentration. Toxicol. Appl. Pharmacol. 74:417-429.

Budavari, S., M.J. O'Neil, A. Smith, P.E. Heckelman, and J.F. Kinneary, eds. 1996. The Merck Index, 12th Ed. Rahway, NJ: Merck & Co., Inc.

C&EN (Chemical & Engineering News). 1994. Chlorine industry running flat out despite persistent health fears. 72:13.

Capodaglio, E., G. Pezzagno, J.C. Bobbio, and F. Cazzoli. 1970. Indagine sulla funzionalita respiratoria di lavoratori addetti a produzioned elettrolitica di chloro e soda [in Italian]. Med. Lav. 60:192-202.

Carlton, B.D., P. Bartlett, A. Basaran, K. Colling, I. Osis, and M.K. Smith. 1986. Reproductive effects of alternative disinfectants. Environ. Health Perspect. 69:237-241.

CEH (Chemical Economics Handbook). 2000. Chlorine/sodium hydroxide [Online]. Available: http://ceh.sric.sri.com/Public/Reports/733.1000/ [October 2001].

Chang, J.C.F., and C.S. Barrow. 1984. Sensory tolerance and cross-tolerance in F-344 rats exposed to chlorine or formaldehyde gas. Toxicol. Appl. Pharmacol. 76:319-327.

Charan, N.B., S. Lakshminarayan, G.C. Myers, and D.D. Smith. 1985. Effects of accidental chlorine inhalation on pulmonary function. West. J. Med. 143:333-336.

CIIT (Chemical Industry Institute of Toxicology). 1993. A Chronic Inhalation Toxicity Study of Chlorine in Female and Male $B6C3F_1$ Mice and Fischer 344 Rats. Research Triangle Park, NC: CIIT.

Cockcroft, D.W., D.N. Killian, J.J.A. Mellon, and F.E. Hargreave. 1977. Bronchial reactivity to inhaled histamine: A method and clinical survey. Clin. Allergy 7:235-243.

D'Alessandro, A., W. Kuschner, H. Wong, H.A. Boushey, and P.D. Blanc. 1996. Exaggerated responses to chlorine inhalation among persons with nonspecific airway hyperreactivity. Chest 109:331-337.

Decker, W.J. 1998. Chlorine poisoning at the swimming pool revisited: Anatomy of two minidisasters. Vet. Hum. Toxicol. 30:584-585.

Demnati, R., R. Fraser, G. Plaa, and J.L. Malo. 1995. Histopathological effects of acute exposure to chlorine gas on Sprague-Dawley rat lungs. J. Environ. Pathol. Toxicol. Oncol. 14:15-19.

Drobnic, F., A. Freixa, P. Casan, J. Sanchis, and X. Guardino. 1996. Assessment of chlorine exposure in swimmers during training. Med. Sci. Sports Exerc. 28:271-274.

Druckrey, H. 1968. Chlorinated drinking water toxicity tests involving seven generations of rats. Food Cosmet. Toxicol. 6:147-154.

DuBois, A.B., S.Y. Botelho, and J.H. Comroe. 1971. A new method for measuring airway resistance in man using a body plethysmograph: Values in normal subjects and in patients with respiratory disease. J. Clin. Invest. 35:327-335.

Eaton, D.L., and C.D. Klaassen. 1996. Principles of toxicology. In Casarett and Doul's Toxicology: The Basic Science of Poisons. New York, NY: McGraw Hill.

Elmes, P.C., and D.P. Bell. 1963. The effects of chlorine gas on the lungs of rats with spontaneous pulmonary disease. J. Pathol. Bacteriol. 86:317-326.

EPA/FEMA/DOT (U.S. Environmental Protection Agency/Federal Emergency Management Association/U.S. Department of Transportation). 1987. Technical Guidance for Hazards Analysis: Emergency Planning for Extremely Hazardous Substances. EPA-OSWER-88-0001. U.S. Environmental Protection Agency, Washington, DC.

EPA (U.S. Environmental Protection Agency). 1994. Methods for Derivation of Inhalation Reference Concentrations and Application of Inhalation Dosimetry. EPA/600/8-90/066F. Environmental Criteria and Assessment Office, Washington, DC.

EPA (U.S. Environmental Protection Agency). 1996. Chlorine. Integrated Risk Information System (IRIS) [Online]. Available: http://www.epa.gov/iris/subst/0405.htm [September 2, 1996].

Ferris, B.G., W.A. Burgess, and J. Worchester. 1964. Prevalence of chronic respiratory disease in a pulp mill and a paper mill in the United States. Br. J. Ind. Med. 24:26-37.

Ferris, B.G., S. Puleo, and H.Y. Chen. 1979. Mortality and morbidity in a pulp and a paper mill in the United States: A ten-year follow-up. Br. J. Ind. Med. 36:127-134.

Freitag. 1941. Dangers of chlorine gas [in German]. Z. Gesamte Schiess-Sprengstoffwesen 35:159.

Gagnaire, F., S. Azim, P. Bonnet, G. Hecht, and M. Hery. 1994. Comparison of the sensory irritation response in mice to chlorine and nitrogen trichloride. J. Appl. Toxicol. 14:405-409.

ILO (International Labour Office). 1998. Chlorine and Compounds. In Encyclopaedia of Occupational Health and Safety, 4th Ed., Vol. 1 (A-K). Geneva, Switzerland: ILO.

Jiang, X.Z., L.A. Buckley, and K.T. Morgan. 1983. Pathology of toxic responses to the RD_{50} concentration of chlorine gas in the nasal passages of rats and mice. Toxicol. Appl. Pharmacol. 71:225-236.

Jones, R.N., J.M. Hughes, H. Glindmeyer, and H. Weill. 1986. Lung function after acute chlorine exposure. Am. Rev. Resp. Dis. 134:1190-1195.

Joosting, P., and M. Verberk. 1974. Emergency population exposure: A methodological approach. Recent Adv. Assess. Health Eff. Environ. Pollut. 4:2005-2029.

Kaufman, J., and B. Burkons. 1971. Clinical, roentgenologic, and physiologic effects of acute chlorine exposure. Arch. Environ. Health 23:29-34.

Klonne, D.R., C.E. Ulrich, M.G. Riley, T.E. Hamm Jr., K.T. Morgan, and C.S. Barrow. 1987. One-year inhalation toxicity study of chlorine in rhesus monkeys (*Macaca mulatta*). Fundam. Appl. Toxicol. 9:557-572.

Kowitz, T.A., R.C. Reba, R.T. Parker, and W.S. Spicer. 1967. Effects of chlorine gas upon respiratory function. Arch. Environ. Health 14:545-558.

Kusch G.D. 1994. Prospective study of the effects of chronic chlorine exposure in manufacturing on respiratory health. In Chlorine Plant Operations Seminar and Workshop Proceedings. Washington, DC: The Chlorine Institute, Inc.

Larsen, G.L. 1992. Asthma in children. N. Engl. J. Med. 326:1540-1545.

Lemiere, C., J.L. Malo, and D. Gautrin. 1996. Nonsensitizing causes of occupational asthma. Obstruct. Lung Dis. 80:749-774.

Lipton, M.A., and G.J. Rotariu. 1941. In Progress report on toxicity of chlorine gas for mice, E.M.K. Gelling and F.C. McLean, eds. Report No. 286. U.S. National Defense Committee, Office of Science Research and Development.

Litchfield, J.T., and F. Wilcoxon. 1949. A simplified method of evaluating dose-effect experiments. J. Pharmacol. Exp. Ther. 96:99.

MacEwen, J.D., and E.H. Vernot. 1972. Toxic Hazards Research Unit Annual Technical Report: 1972. AMRL-TR-72-62. Aerospace Medical Research Laboratory, Wright-Patterson Air Force Base, OH; National Technical Information Service, Springfield, VA.

Matheson Gas Co. 1980. Matheson Gas Data Book, 6th Ed. Lyndhurst, NJ: Division Searle Medical Products USA, Inc.

Meier, J.R., R.J. Bull, J.A. Stober, and M.C. Cimino. 1985. Evaluation of chemicals used for drinking water disinfection for production of chromosomal damage and sperm-head abnormalities in mice. Environ. Mut. 7:201-211.

Mickey, G., and H. Holden. 1971. Effects of chlorine on mammalian cells in vitro. EMS Newsletter 4:39-41.

Mrvos, R., B.S. Dean, and E.P. Krenzelok. 1993. Home Exposures to chlorine/chloramine gas: Review of 216 cases. South. Med. J. 86:654-657.

NAC (National Advisory Committee). 1996. Acute Exposure Guideline Levels (AEGLs) for Fluorine. Interim Report.

NIOSH (National Institute for Occupational Safety and Health). 1976. Criteria for a recommended standard: Occupational exposure to chlorine. NIOSH Publication 76-170. U.S. Department of Health and Human Services, Washington, DC.

NIOSH (National Institute for Occupational Safety and Health). 1994a. Documentation for Immediately Dangerous to Life or Health Concentrations (IDLHs). PB94-195047. National Institute for Occupational Safety and Health, Cincinnati, OH; National Technical Information Service, Springfield, VA.

NIOSH (National Institute for Occupational Safety and Health). 1994b. NIOSH Pocket Guide to Chemical Hazards. NIOSH Publication 94-116. U.S. Department of Health and Human Services, U.S. Government Printing Office, Washington, DC.

NRC (National Research Council). 1984. Emergency and Continuous Exposure Limits for Selected Airborne Contaminants, Vol. 2. Washington, DC: National Academy Press.

NRC (National Research Council). 1991. Permissible Exposure Levels and Emergency Exposure Guidance Levels for Selected Airborne Contaminants. Washington, DC: National Academy Press.

NRC (National Research Council). 1993. Guidelines for Developing Community Emergency Exposure Levels for Hazardous Substances. Washington, DC: National Academy Press.

NRC (National Research Council). 2001. Standing Operating Procedures for Developing Acute Exposure Guideline Levels for Hazardous Chemicals. Washington, DC: National Academy Press.

NTIS (National Technical Information Service). 1996. Chlorine. Registry of Toxic Effects of Chemical Substances (RTECS) [Online]. Available: Available: http://www.ntis.gov/search/product.asp?ABBR=SUB5363&starDB =GRAHIST [December 1996].

NTP (National Toxicology Program). 1992. Toxicology and Carcinogenesis of Chlorinated Water (CAS Nos. 7782-50-5 and 7681-52-9) and Chloraminated Water (CAS No. 10599-90-3) (Deionized and Charcoal-Filtered) in F344/N Rats and $B6C3F_1$ Mice (Drinking Water Studies). NTP TR 392. National Institutes of Health, Bethesda, MD.

O'Neil, C.E. 1991. Immune responsiveness in chlorine exposed rats. PB92-124478. National Institute for Occupational Safety and Health, Cincinnati, OH.

OSHA (Occupational Safety and Health Administration). 1989. 29 CFR Part 1910, Air Contaminants: Final Rule. Fed. Regist. 54(12):2455-2456 (Thursday, January 19, 1989).

Patil, L.R.S., R.G Smith, A.J. Vorwald, and T.F. Mooney. 1970. The health of diaphragm cell workers exposed to chlorine. Am. Ind. Hyg. Assoc. J. 31:678-686.

Perry, W.G., F.A. Smith, and M.B. Kent. 1994. The halogens. Pp. 4482-4505 in Patty's Industrial Hygiene and Toxicology, Vol. 2, Part F, G.F. Clayton and F.E. Clayton, eds. New York, NY: John Wiley & Sons, Inc.

Prentiss, A.M. 1937. Chemicals in War; a Treatise on Chemical Warfare. New York: McGraw-Hill Book Company.

Rafferty, P. 1980. Voluntary chlorine inhalation: A new form of self-abuse? Br. Med. J. 281:1178-1179.

Rothery, S.P. 1991. Hazards of chlorine to asthmatic patients. Br. J. Gen. Pract. 41:39.

Rotman, H.H., M.J. Fliegelman, T. Moore, R.G. Smith, D.M. Anglen, C.J. Kowalski, and J.G. Weg. 1983. Effects of low concentration of chlorine on pulmonary function in humans. J. Appl. Physiol. 54:1120-1124.

Rupp, H., and D. Henschler. 1967. Effects of low chlorine and bromine concentrations on man. Int. Arch. Gewerbepathol. 23:79-90.

Schlagbauer, M., and D. Henschler. 1967. Toxicity of chlorine and bromine with single and repeated inhalation. Int. Arch. Gewerbepath. Gewerbehyg. 23:91.

Shroff, C.P., M.V. Khade, and M. Srinivasan. 1988. Respiratory cytopathology in chlorine gas toxicity: A study in 28 subjects. Diagn. Cytopathol. 4:28-32.

Shusterman, D.J., M.A. Murphy, and J.R. Balmes. 1998. Subjects with seasonal allergic rhinitis and nonrhinitic subjects react differentially to nasal provocation with chlorine gas. J. Allergy Clin. Immunol. 101:732-740.

Silver, S.D., F.P. McGrath, and R.L. Ferguson. 1942. Chlorine median lethal concentration data for mice. DATR 373. Edgewood Arsenal, MD.

ten Berge, W.F., and M. Vis van Heemst. 1983. Validity and accuracy of a commonly used toxicity-assessment model in risk analysis. IChemE Symposium Series No. 80:I1-I12.

ten Berge, W.F., A. Zwart, and L.M. Appleman. 1986. Concentration-time mortality response relationship of irritant and systemically acting vapors and gases. J. Hazard. Mater. 13:301-310.

Underhill, F.P. 1920. The lethal War Gases: Physiology and Experimental Treatment. New Haven, CT: Yale University Press. Pp. 20.

Vernot, E.H., J.D. MacEwen, C.C. Haun, and E.R. Kinkead. 1977. Acute toxicity and skin corrosion data for some organic and inorganic compounds and aqueous solutions. Toxicol. Appl. Pharmacol. 42:417-423.

Weedon, F.R, A. Hartzell, and C. Setterstrom. 1940. Toxicity of ammonia, chlorine, hydrogen cyanide, hydrogen sulfide and sulfur dioxide gases. V. Animals. Contrib. Boyce Thompson Inst. 11:365-385.

Weill, H., R. George, M. Schwarz, and M. Ziskind. 1969. Late evaluation of pulmonary function after acute exposure to chlorine gas. Amer. Rev. Resp. Dis. 99:374-379.

Withers, R.M.J., and F.P. Lees. 1985a. The assessment of major hazards: The lethal toxicity of chlorine, Part 1: Review of information on toxicity. J. Hazard. Mater. 12:231-282.

Withers, R.M.J., and F.P. Lees. 1985b. The assessment of major hazards: The lethal toxicity of chlorine., Part 2: model of toxicity to man. J. Hazard. Mater. 12:283-302.

Withers, R.M.J., and F.P. Lees. 1987. The assessment of major hazards: The lethal toxicity of chlorine, Part 3: Crosschecks from gas warfare. J. Hazard. Mater. 15:301-342.

Witschi, H.R., and J.A. Last. 1996. Toxic responses of the respiratory system. In Casarett and Doull's Toxicology: The Basic Science of Poisons. New York, NY: McGraw-Hill.

Wohlslagel, J., L.C. DiPasquale, and E.H. Vernot. 1976. Toxicity of solid rocket motor exhaust: Effects of HCl, HF, and alumina on rodents. J. Combust. Toxicol. 3:61-69.

Wolf, D.C., K.T. Morgan, E.A. Gross, C. Barrow, O.R. Moss, R.A. James, and J.A. Popp. 1995. Two-year inhalation exposure of female and male B6C3F1 mice and F344 rats to chlorine gas induces lesions confined to the nose. Fundam. Appl. Toxicol. 24:111-131.

Wood, B.R., J.L. Colombo, and B.E. Benson. 1987. Chlorine inhalation toxicity from vapors generated by swimming pool chlorinator tablets. Pediatrics 79:427-430.

Zwart, A., and R.A. Woutersen. 1988. Acute inhalation toxicity of chlorine in rats and mice: Time-concentration-mortality relationships and effects on respiration. J. Hazard. Mater. 19:195-208.

APPENDIX A

Derivation of Chlorine AEGLs

Derivation of AEGL-1

Key studies: Rotman et al. 1983; D'Alessandro et al. 1996; Shusterman et al. 1998

Toxicity end point: Transient pulmonary function changes in atopic individual exposed at 0.5 ppm for an interrupted 8 h; non-significant changes in pulmonary peak air flow in eight atopic individuals exposed at 0.5 ppm for 15 min; no statistically significant pulmonary parameter changes in asthmatic subjects exposed at 0.4 ppm for 1 h

Time-scaling: No time scaling; because there is adaptation to the slight irritation that defines the AEGL-1 end point, the same value (0.5 ppm) was used across all time points

Uncertainty factors: 1, because susceptible individuals were tested and one of the susceptible individuals was exercising, making him more susceptible to sensory irritation (no-effect level in healthy exercising individuals of both genders)

Calculations: Because the 0.5 ppm concentration was indicative of a NOAEL for more serious pulmonary changes, the 0.5 ppm concentration was used for all exposure durations. The susceptible individual underwent an interrupted 8-h exposure at 0.5 ppm without increased symptoms, so that concentration was also used for the 8-h AEGL-1

Derivation of AEGL-2

Key studies: Rotman et al. 1983; D'Alessandro et al. 1996

Toxicity end point: No-effect concentration for serious health effect (asthma-like attack) in a sensitive, exercising individual exposed at 1 ppm for 4 h and in individuals with airway hyper-reactivity (including 3 asthmatic individuals) exposed at 1 ppm for 1 h

Time-scaling: $C^2 \times t = k$ (ten Berge and Vis van Heemst 1983)
$(1\ \text{ppm})^2 \times 240\ \text{min} = 240\ \text{ppm}^2\cdot\text{min}$

Uncertainty factors: 1. The value was based on effects consistent with the AEGL-2 definition in a susceptible, exercising individual and in asthmatics subjects

30-min AEGL-2: $C^2 \times 30\ \text{min} = 240\ \text{ppm}^2\cdot\text{min}$
$C = 2.8$ ppm

1-h AEGL-2: $C^2 \times 60\ \text{minutes} = 240\ \text{ppm}^2\cdot\text{min}$
$C = 2$ ppm

4-h AEGL-2: 1 ppm for 4 h; basis for derivation of other exposure durations

8-h AEGL-2: $C^2 \times 480\ \text{min} = 240\ \text{ppm}^2\cdot\text{min}$
$C = 0.71$ ppm

The 10-min AEGL-2 was set equal to the 30-min AEGL-2 so that the highest human test concentration of 4.0 ppm was not exceeded.

Derivation of AEGL-3

Key studies: Zwart and Woutersen 1988; MacEwen and Vernot 1972

Toxicity end point: 1-h lethality value; an end point below the highest concentration resulting in no deaths in the rat and above the highest concentration resulting in no deaths in the mouse was chosen because the mouse was shown to be more sensitive than other mammals to irritant gases, including chlorine, and does not provide an appropriate basis for quantitatively predicting mortality in humans

Time-scaling: $C^2 \times t = k$ (ten Berge and Vis van Heemst 1983)
$(200 \text{ ppm}/10)^2 \times 60 \text{ min} = 24{,}000 \text{ ppm}^2\cdot\text{min}$

Uncertainty factors: Combined uncertainty factor of 10

3 for interspecies variability (interspecies values for the same end point differed by a factor of approximately 2 in several studies)

3 for differences in human sensitivity (the toxic effect is the result of a chemical reaction with biologic tissue of the respiratory tract, which is unlikely to differ among individuals)

10-min AEGL-3: $C^2 \times 10 \text{ min} = 24{,}000 \text{ ppm}^2\cdot\text{min}$
$C = 50$ ppm

30-min AEGL-3: $C^2 \times 30 \text{ min} = 24{,}000 \text{ ppm}^2\cdot\text{min}$
$C = 28.3$ ppm

1-h AEGL-3: 200 ppm/10 = 20 ppm

4-h AEGL-3: $C^2 \times 240 \text{ min} = 24{,}000 \text{ ppm}^2\cdot\text{min}$
$C = 10$ ppm

8-h AEGL-3: $C^2 \times 480 \text{ min} = 24{,}000 \text{ ppm}^2\cdot\text{min}$
$C = 7.1$ ppm

APPENDIX B

ACUTE EXPOSURE GUIDELINE LEVELS FOR CHLORINE (CAS No. 7782-50-5)

DERIVATION SUMMARY

AEGL-1				
10 min	30 min	1 h	4 h	8 h
0.5 ppm	0.5 ppm	0.5 ppm	0.50 ppm	0.50 ppm
Key references: (1) Rotman, H.H., M.J. Fliegelman, T. Moore, R.G. Smith, D.M. Anglen, C.J. Kowalski, and J.G. Weg. 1983. Effects of low concentrations of chlorine on pulmonary function in humans. J. Appl. Physiol. 54:1120-1124. (2) Shusterman, D.J., M.A. Murphy, and J.R. Balmes. 1998. Subjects with seasonal allergic rhinitis and nonrhinitic subjects react differentially to nasal provocation with chlorine gas. J. Allergy Clin. Immunol. 101:732-740. (3) D'Alessandro, A., W. Kuschner, H. Wong, H.A. Boushey, and P.D. Blanc. 1996. Exaggerated responses to chlorine inhalation among persons with nonspecific airway hyperreactivity. Chest 109:331-337.				
Test species/strain/number: Eight male subjects, one atopic subject (Rotman et al. 1983); eight atopic subjects and eight nonatopic subjects (Shusterman et al. 1998); five asthmatic subjects and five nonasthmatic subjects (D'Alessandro et al. 1996).				
Exposure route/concentrations/durations: Inhalation; 0.0, 0.5, 1.0 ppm for 8 h; break at 4 h for an unreported period of time to undergo pulmonary function tests followed by chamber reentry; subjects exercised for 15 min of every hour during exposures; sham exposures were included (Rotman et al. 1983) Inhalation; 0.0 ppm or 0.5 ppm for 15 min (Shusterman et al. 1998) Inhalation; 0.4 ppm or 1.0 ppm for 1 h (D'Alessandro et al. 1996)				
Effects: 0.5 ppm for 4 h—no effects in eight of nine subjects; transient changes in pulmonary functions in one of nine subjects. 1.0 ppm for 4 h—some irritation, transient changes in pulmonary functions in nine subjects including an atopic individual; asthma-like episode in one of nine subjects when exposure duration extended to more than 4 h (Rotman et al. 1983).				

AEGL-1 *Continued*
0.5 ppm for 15 min—nasal congestion; nonsignificant changes in pulmonary peak flow (Shusterman et al. 1998). 0.4 ppm for 1 h—no statistically significant pulmonary function effect in asthmatic individuals (D'Alessandro et al. 1996).
End point/concentration/rationale: 0.5 ppm forr 4 h resulted in no effects in healthy human subjects and transient changes in pulmonary functions for a susceptible individual who had obstructive airways disease prior to the exposure. The 0.5-ppm concentration was chosen as the basis for the AEGL-1 because the next highest concentration produced effects consistent with an AEGL-2 (coughing, wheezing, and a considerable increase in airways resistance) in a susceptible individual. Supported by studies of Shusterman et al. (1998) and D'Alessandro et al. (1996).
Uncertainty factors/rationale: Total uncertainty factor: 1 Interspecies: Not applicable; human subjects tested. Intraspecies:1. An atopic individual who had obstructive airways disease prior to the exposure and was considered characteristic of the "susceptible" population was tested. This individual was did not exhibit adverse effects. The choice of an intraspecies uncertainty factor of 1 is supported by another study in which a concentration of 0.4 ppm for 1 h was a no-effect concentration for changes in pulmonary function parameters in individuals with airway hyper-reactivity/asthma and by a study in asthmatic subjects exposed at 0.4 ppm
Modifying factor: Not applicable
Animal to human dosimetric adjustment: Not applicable; human data used
Time-scaling: Not applied; because 0.5 ppm appeared to be the threshold for more severe changes in pulmonary parameters in the atopic individual regardless of exposure duration, the 0.5 ppm was used for all AEGL-1 exposure durations.
Data adequacy: The Angelen (1981) study was well conducted and documented and reinforces a study conducted earlier at the same facilities in which 31 male and female subjects were tested for sensory irritation. The Rotman et al. (1983) study went into greater detail than the earlier study, measuring 15 pulmonary function parameters before, during, and after exposures. Subjects were exercising during exposures and this study included a susceptible individual. The choice of intraspecies uncertainty factor was supported by a study of shorter duration with asthmatics.

AEGL-2				
10 min	30 min	1 h	4 h	8 h
2.8 ppm	2.8 ppm	2.0 ppm	1.0 ppm	0.71 ppm
Key references: (1) Rotman, H.H., M.J. Fliegelman, T. Moore, R.G. Smith, D.M. Anglen, C.J. Kowalski, and J.G. Weg. 1983. Effects of low concentrations of chlorine on pulmonary function in humans. J. Appl. Physiol. 54:1120-1124. (2) D'Alessandro, A., W. Kuschner, H. Wong, H.A. Boushey, and P.D. Blanc. 1996. Exaggerated responses to chlorine inhalation among persons with nonspecific airway hyperreactivity. Chest 109:331-337.				
Test species/strain/gender/number: Nine human male subjects, including atopic individual (Rotman et al. 1983); 10 human subjects of which five had airway reactivity/asthma (D'Alessandro et al. 1996)				
Exposure route/concentration/duration: Inhalation; 0.0, 0.5, 1.0 ppm for 8 h; break at 4 h for an unreported time period to undergo pulmonary function tests followed by chamber reentry; subjects exercised for 15 min of every hour during exposures; sham exposures were included (Rotman et al. 1983) Inhalation; 0.4 or 1.0 ppm for 1 h (D'Alessandro et al. 1996)				
Effects: 0.5 ppm for 4 h—no effects in eight healthy subjects; transient changes in pulmonary functions in one of nine subjects. 1.0 ppm for 4 h—some irritation, transient changes in pulmonary functions nine subjects including an atopic individual; asthma-like episode in one of nine subjects when exposure duration extended beyond 4 h. 1.0 ppm for 1 h—increased airway resistance in asthmatic individuals (D'Alessandro et al. 1996).				
End point/concentration/rationale: 1 ppm for 4 h was a no-effect exposure for serious health symptoms in an atopic exercising individual, and 1 ppm for 1 h was a symptomless effect on airway resistance in asthmatic individuals. However, the increase in airways resistance was considered the NOAEL for an AEGL-2 effect.				
Uncertainty factors/rationale: Total uncertainty factor: 1 Interspecies: Not applicable; human subjects tested. Intraspecies: 1. A susceptible exercising individual who had obstructive airways disease prior to the exposure and was considered characteristic of the "susceptible" population was tested. The application of an intraspecies uncertainty factor of 1 is supported by another study in which individuals with airway hyperreactivity/asthma				

AEGL-2 *Continued*
showed similar pulmonary function changes and some clinical symptoms but no asthma-like attack following exposure at 1.0 ppm for 1 h.
Modifying factor: Not applicable
Animal to human dosimetric adjustment: Not applicable; human data used
Time-scaling: $C^n \times t = k$ where $n = 2$. This value describes the concentration-exposure duration relationship for the end point of nuisance irritation (ten Berge and Vis van Heemst 1983, IChemE Symposium Series No. 80:17-21).
Data adequacy: The Angelen (1981) study was well conducted and documented and reinforces a study conducted earlier at the same facilities in which 31 male and female subjects were tested for sensory irritation. The Rotman et al. (1983) study went into greater detail than the earlier study, measuring 15 pulmonary function parameters before, during, and after exposures. Subjects were exercising during exposures, and a susceptible individual was included. The choice of intraspecies uncertainty factor was supported by a study of shorter duration with asthmatic subjects (D'Alessandro et al. 1996).

AEGL-3				
10 min	30 min	1 h	4 h	8 h
50 ppm	28 ppm	20 ppm	10 ppm	7.1 ppm
Key references: (1) MacEwen, J.D. and E.H. Vernot. 1972, Toxic Hazards Research Unit Annual Technical Report. 1972. Wright-Patterson Air Force Base, Dayton, OH; (2) Zwart, A. and Woutersen. 1988. Acute inhalation toxicity of chlorine in rats and mice: time-concentration mortality relationships and effects on respiration. J. Hazard. Mater. 19:195-208; (3) O'Neil, C.E. 1991. Immune responsiveness in chlorine exposed mice. PB92-124478, Prepared for NIOSH, Cincinnati, OH.				
Test species/strain/gender/number: (1) Sprague-Dawley rats, 10/exposure group; (2) Wistar-derived rats, 10/exposure group; (3) BALB/c mice, 10/exposure group				
Exposure route/concentrations/durations: Inhalation; (1) 213-427 ppm for 1 h, (2) 322-595 ppm for 1 h, (3) 50-250 ppm for 1 h				
Effects: (1) no deaths at 213 ppm for 1 h (Sprague-Dawley rat); (2) no deaths at 322 ppm for 1 h (Wistar-derived rat); (3) no deaths at 150 ppm for 1 h (BALB/c mouse)				
End point/concentration/rationale: 200 ppm for 1 h (the estimated mean of highest experimental nonlethal values for the rat and mouse) was chosen as the basis for the 1-h AEGL-3. Mice appeared to be unusually sensitive to chlorine, and in some studies, delayed deaths were attributed to bronchopneumonia rather than direct effects of chlorine.				
Uncertainty factors/rationale: Total uncertainty factor: 10 Interspecies: 3. The mouse and rat LC_{50} values did not differ by more than a factor of 2 to 3, and the mouse was consistently more sensitive. In some mouse studies delayed deaths were attributed to bronchopneumonia rather than direct effects of chlorine exposure. Intraspecies: 3. Chlorine is a highly reactive, irritating, and corrosive gas whose effect on respiratory tissues is not expected to differ greatly among individuals.				
Modifying factor: Not applicable				
Animal to human dosimetric adjustment: Not applied				
Time-scaling: $C^n \times t = k$ where $n = 2$. This value describes the concentration-exposure duration relationship for the end point of nuisance irritation (ten Berge and Vis van Heemst 1983, IChemE Symposium Series				

AEGL-3 *Continued*
No. 80:17-21). The irritation mechanism of action leads to pulmonary edema and potential lethality. An *n* of 2 is also relevant to animal lethality studies.
Data adequacy: The database for chlorine is extensive with multiple studies of lethality conducted at several exposure durations and involving several species. Studies with multiple dosing regimens showed a clear dose-response relationship. Longer-term studies that support the safety of the values were also available. Tissue and organ pathology indicated that the toxic mechanism was the same across species.

2

Hydrogen Chloride[1]

Acute Exposure Guideline Levels

SUMMARY

Hydrogen chloride (HCl) is a colorless gas with a pungent, suffocating odor. It is used in the manufacture of organic and inorganic chemicals, oil-well acidizing, steel pickling, food processing, and minerals and metals processing. A large amount of HCl is released from solid rocket fuel exhaust. It is an upper respiratory irritant at relatively low concentrations and may cause damage to the lower respiratory tract at higher concentrations. HCl is very soluble in water, and the aqueous solution is highly corrosive.

---

[1]This document was prepared by the AEGL Development Team comprising Cheryl Bast (Oak Ridge National Laboratory) and National Advisory Committee (NAC) for Acute Exposure Guideline Levels for Hazardous Substances member John Hinz (Chemical Manager). The NAC reviewed and revised the document and the AEGL values as deemed necessary. Both the document and the values were then reviewed by the National Research Council (NRC) Subcommittee on Acute Exposure Guideline Levels. The NRC subcommittee concluded that the AEGLs developed in this document are scientifically valid conclusions on the basis of the data reviewed by the NRC and are consistent wit the NRC guidelines reports (NRC 1993, 2001).

The lowest acute exposure guideline level (AEGL) values are based on a 45-minute (min) no-observed-adverse- effect level (NOAEL) of 1.8 parts per million (ppm) in exercising adult asthma patients (Stevens et al. 1992). No uncertainty factors (UFs) were applied for inter- or intraspecies variability because the study population consisted of sensitive humans. The same 1.8-ppm value was applied across the 10- and 30-min and 1-, 4-, and 8-hour (h) exposure times, because mild irritance generally does not vary greatly over time, and because it is not expected that prolonged exposure will result in an enhanced effect.

The AEGL-2 for the 30-min and 1- , 4- , and 8-h time points was based on severe nasal or pulmonary histopathology in rats exposed at 1,300 ppm for 30 min (Stavert et al. 1991). A modifying factor (MF) of 3 was applied to account for the relatively sparse database describing effects defined by AEGL 2. The AEGL-2 values were further adjusted by a total UF of 10—3 for intraspecies variability, supported by the steep concentration- response curve, which implies little individual variability; and 3 for interspecies variability. Using the default value of 10 for interspecies variability would bring the total adjustment to 100 (total UF × MF) instead of 30. That would generate AEGL-2 values that are not supported by the total data set, including data on exercising asthmatic subjects, an especially sensitive subpopulation, because exercise increases HCl uptake and exacerbates irritation; no effects were noted in exercising young adult asthmatic subjects exposed to HCl at 1.8 ppm for 45 min (Stevens et al. 1992). A total UF of 10, accompanied by the MF of 3, is most consistent with the total database (see Section 6.3 for detailed support of uncertainty factors). Thus, the total factor is 30. Time-scaling for the 1-h AEGL exposure period was accomplished using the $C^n \times t = k$ relationship (C = concentration, t = time, and k is a constant), where $n = 1$ based on regression analysis of combined rat and mouse LC_{50} data (concentrations lethal to 50% of subjects) (1 min to 100 min) as reported by ten Berge et al. (1986). The 4- and 8-h AEGL-2 values were derived by applying an MF of 2 to the 1-h AEGL-2 value, because time-scaling would yield a 4-h AEGL-2 of 5.4 ppm and an 8-h AEGL-2 of 2.7 ppm, close to the 1.8 ppm tolerated by exercising asthmatic subjects without adverse health effects. The 10-min AEGL 2 was derived by dividing the mouse RD_{50} (concentration expected to cause a 50% decrease in respiratory rate) of 309 ppm by a factor of 3 to obtain a concentration causing irritation (Barrow 1977). It has been determined that human response to sensory irritants can be predicted on the basis of the mouse RD_{50}. For example, Schaper (1993) has validated the correlation of $0.03 \times RD_{50}$ =

TABLE 2-1 Summary of AEGLs Values for Hydrogen Chloride (ppm [mg/m³])

Classification	10 min	30 min	1 h	4 h	8 h	End Point (Reference)
AEGL-1 (Nondisabling)	1.8 (2.7)	1.8 (2.7)	1.8 (2.7)	1.8 (2.7)	1.8 (2.7)	NOAEL in exercising asthmatic subjects (Stevens et al. 1992)
AEGL-2 (Disabling)	100 (156)	43 (65)	22 (33)	11 (17)	11 (17)	Mouse RD_{50} (Barrow et al. 1977); histopathology in rats (Stavert et al. 1991)
AEGL-3 (Lethal)	620 (937)	210 (313)	100 (155)	26 (39)	26 (39)	Estimated NOEL for death from 1-h rat LC_{50} (Wohlslagel et al. 1976; Vernot et al. 1977)

Abbreviations: LC_{50}, concentration lethal to 50% of subjects; mg/m³, milligrams per cubic meter; NOAEL, no-observed-adverse-effect level; NOEL, no-observed-effect level; ppm, parts per million; RD_{50}, concentration expected to cause a 50% decrease in respiratory rate.

Threshold Limit Value (TLV) as a value that will prevent sensory irritation in humans. The multiplier 0.03 represents the half-way point between 0.1 and 0.01 on a logarithmic scale, and Alarie (1981) has shown that the RD_{50} multiplied by 0.1 corresponds to "some sensory irritation," while the RD_{50} value itself is considered "intolerable to humans." Thus, it is reasonable that one third of the RD_{50}, a value half-way between 0.1 and 1 on a logarithmic scale, may cause significant irritation to humans. Furthermore, one-third of the mouse RD_{50} for HCl corresponds to an approximate decrease in respiratory rate of 30%, and decreases in the range of 20-50% correspond to moderate irritation (ASTM 1991).

The AEGL-3 values were based on a 1-h rat LC_{50} study (Wohlslagel et al. 1976; Vernot et al. 1977). One-third of the 1-h LC_{50} of 3,124 ppm was used to estimate a concentration causing no deaths. That estimate is inherently conservative (no deaths were observed in the same study at 1,813 ppm). A total UF of 10 will be applied—3 for intraspecies variation, because the steep concentration-response curve implies limited individual variability; and 3 to protect susceptible individuals. Using a full value of 10 for interspecies variability (total UF of 30) would yield AEGL-3 values that are inconsistent with the overall data set (see Section 7.3 for detailed support of UFs). Thus, the total UF is 10. The value was then time- scaled to the specified 10- and 30- min and 4-h AEGL exposure periods using the $C^n \times t = k$ relationship, where $n = 1$ based on regression analysis of combined rat and mouse LC_{50} data (1 min to 100 min) as reported by ten Berge et al. (1986). The 4-h AEGL-3 value was also adopted as the 8-h AEGL-3 value because of the added uncertainty of time scaling to 8-h using a value of n derived for exposure durations up to 100 min.

1. INTRODUCTION

HCl is a colorless gas with a pungent, suffocating odor. It is hygroscopic and produces whitish fumes in moist air. HCl is produced as a byproduct of chemical syntheses of chlorinated compounds and is used in the manufacture of organic and inorganic chemicals, oil-well acidizing, steel pickling, food processing, and the processing of minerals and metals. A large amount of HCl is released from solid rocket fuel exhaust. It is very soluble in water, and the aqueous hydrochloric acid is quite corrosive (EPA 1994). The physicochemical data for hydrogen chloride are shown in Table 2-2.

2. HUMAN TOXICITY DATA

2.1. Acute Lethality

2.1.1. Case Reports

No data concerning human lethality from HCl exposure were located in the available literature.

TABLE 2-2 Physicochemical Data for Hydrogen Chloride

Parameter	Value	Reference
Synonyms	Muriatic acid, hydrochloric acid	AIHA 1989
Chemical formula	HCl	AIHA 1989
Molecular weight	36.47	AIHA 1989
CAS registry no.	7647-01-0	AIHA 1989
Physical state	Colorless, fuming gas	AIHA 1989
Relative density	1.268 at 25°C	AIHA 1989
Boiling/flash point	-85°C/nonflammable	AIHA 1989
Solubility in water	Very soluble (82.3 g/100 ml)	EPA 1994
Conversion factors in air	1 mg/m^3 = 0.67 ppm 1 ppm = 1.49 mg/m^3	AIHA 1989

2.2. Nonlethal Toxicity

2.2.1. Experimental Studies

Five male and five female adult asthmatic subjects (age 18 to 25 years [y]) were exposed to filtered air or HCl at 0.8 ppm or 1.8 ppm for 45 min (Stevens et al. 1992). Exposure levels were verified by an online filtering system during exposures and analyzed by ion exchange chromatography. Actual mean exposure concentrations were 0, 0.8 ± 0.09, or 1.84 ± 0.21 ppm. The subjects were healthy, except for having asthma, and wore half-face masks to allow for nasal and oral breathing and to control exposure of the eyes. The 45-min exposure sessions consisted of 15 min of exercise (treadmill walking at 2 miles per hour at an elevation grade of 10%) followed by 15 min of rest followed by another 15 min of exercise. Exposures to the test atmospheres were separated by at least 1 week (wk). Subjects rated severity of symptoms before, during, and after exposure on a scale of 1 to 5 (5 being most severe). Symptoms rated included upper respiratory (sore throat and nasal discharge), lower respiratory (cough, chest pain or burning, dyspnea, wheezing), and other (fatigue, headache, dizziness, unusual taste or smell). Pulmonary function measurements were performed while subjects were seated in a pressure-compensated volume-displacement body plethysmograph. The following parameters were measured: total respiratory resistance, thoracic gas volume at functional residual capacity, forced expiratory volume, forced vital capacity, and maximal flow at 50% and 75% of expired vital capacity. Nasal work of breathing and oral ammonia levels were also measured. No adverse treatment-related effects were observed. There were no treatment-related increases in severity of upper respiratory, lower respiratory, or other symptoms reported by participants. No significant differences were reported between test and control exposures with regard to any of the pulmonary function tests. No treatment-related changes were observed in nasal work of breathing data. Oral ammonia levels showed a significant increase after exposure to both concentrations of HCl but not after exposure to air; the study authors conclude that this finding is counterintuitive and offer no explanation for the observation.

2.2.2. Case Reports

Reactive airways dysfunction syndrome (RADS) is an asthma-like condition that develops after a single exposure to high levels of a chemical irritant. Symptoms occur within minutes to hours after the initial exposure and may persist as nonspecific bronchial hyper-responsiveness for months to years (Bernstein 1993). It was given the name RADS by Brooks et al. (1985) in a retrospective analysis of 10 previously healthy people who had developed persistent airway hyper-reactivity after a single, high-level exposure to a chemical irritant. The acronym then gained acceptance in the medical community (Nemery 1996), because a name had finally been given to a clinical entity that physicians had encountered. Little or no published evidence had been previously available to verify the claim that asthma symptoms could be a consequence of a single inhalation exposure. This syndrome has been described after exposure to HCl. Promisloff et al. (1990) reported RADS in three male police officers (36-45 y old) who responded to a roadside chemical spill. The subjects were exposed to unquantified amounts of sodium hydroxide, silicon tetrachloride, and HCl as a by-product of trichlorosilane hydrolysis; due to the mixture if irritants involved in the release, it is likely that all compounds contributed to the RADS observed after this accident. In another report, Boulet (1988) described the case of a 41-y-old male nonsmoker who had a 6-y history of mild asthma. After cleaning a pool for 1 h with a solution containing hydrochloric acid, he developed a rapidly progressive and severe bronchospasm that was eventually diagnosed as RADS. No exposure concentration was reported. Turlo and Broder (1989) describe a retrospective review of occupational asthma records. A 57-y-old male, with a smoking history of

12 pack-years, had symptoms consistent with RADS after occupational exposure to hydrochloric acid and phosgene. No exposure concentrations were reported.

Other data concerning acute inhalation exposure to HCl in humans are qualitative and dated, making accurate exposure assessment difficult. A summary of those data is presented in Table 2-3.

2.2.3. Epidemiologic Studies

Epidemiologic studies regarding human exposure to HCl were not available.

2.3. Developmental and Reproductive Toxicity

No human developmental or reproductive toxicity data concerning HCl were identified in the available literature.

2.4. Genetic Toxicology

No data concerning the genotoxicity of HCl in humans were identified in the available literature.

2.5. Carcinogenicity

Data concerning carcinogenicity from exposure to HCl are equivocal. A study of U.S. steel-pickling workers showed an excess risk for lung cancer in individuals exposed primarily to hydrochloric acid for at least 6 months (mo) (Beaumont et al. 1987, as cited in IARC 1992). However, no exposure concentrations were available and the subjects had also been exposed to mists of other acids. In a follow-up of the same cohort, Steenland et al. (1988) observed an excess incidence of laryngeal cancer. Again, the data are confounded by possible exposure to other acid gases, including sulfuric acid. In three case-control studies, no association was observed between occupational exposure to HCl and lung (Bond et al. 1986, as cited in IARC 1992), brain (Bond et al. 1983), or kidney (Bond et al. 1985) cancer. In another report, Bond et al. (1991) examined the records of 308 workers who

TABLE 2-3 Inhalation Exposure of Humans to Hydrogen Chloride

Approximate Concentration	Exposure Time	Effect	Reference
0.77 ppm; 1-5 ppm; 10 ppm	Unspecified	Geometric mean of odor thresholds; odor threshold; odor threshold	Amoore and Hautala 1983; Heyroth 1963; Leonard et al. 1969
0.8 ppm and 1.8 ppm	45 min	No effects in exercising asthmatic subjects	Stevens et al. 1992
≥5 ppm	Unspecified	Immediately irritating	Elkins 1959
>10 ppm	Occupational	Highly irritating, although workers develop some tolerance	Elkins 1959
10 ppm	Prolonged	Maximum tolerable	Henderson and Hagard 1943
10-50 ppm	A few hours	Maximum tolerable	Henderson and Hagard 1943
35 ppm	Short	Throat irritation	Henderson and Hagard 1943
50-100 ppm	1 h	Maximum tolerable	Henderson and Hagard 1943
1,000-2,000 ppm	Short	Dangerous	Henderson and Hagard 1943

died of lung, bronchus, or trachea cancer. The workers were divided into groups for exposure duration as follows: <1 y, 1-4.9 y, or >5 y. Exposure concentrations were 0, 0.25, 1.5, or 3.75 ppm. No association was found between HCl exposure and cancer incidence. In a Canadian population-based case-control study, an increased risk for oat cell carcinoma was suggested in workers exposed to hydrochloric acid; however, no excess risk was observed for all types of lung cancer combined or for other histological types of lung cancer individually (Siemiatycki 1991, as cited in IARC 1992).

2.6. Summary

No treatment-related effects were observed in exercising, young adult asthmatic subjects exposed to HCl at 0.8 ppm or 1.8 ppm for 45 min. Reactive airway dysfunction syndrome (RADS) has been described in people exposed to undetermined concentrations of HCl. Data concerning carcinogenicity from exposure to HCl are equivocal and are confounded by occupational exposure to other chemicals. No data concerning genetic toxicology or developmental or reproductive toxicity in humans from HCl exposure were located in the available literature.

3. ANIMAL TOXICITY DATA

3.1. Acute Lethality

3.1.1. Guinea Pigs

Malek and Alarie (1989) observed 100% mortality in exercising guinea pigs exposed to HCl at 586 ppm for approximately 3 min, although no deaths were observed in guinea pigs exposed at 162 ppm for 30 min. Burleigh- Flayer et al. (1985) exposed guinea pigs to HCl at 320, 680, 1,040, or 1,380 ppm for 30 min. Mortality was as follows: 2/8 during exposure at 1,380 ppm; 1/8 following exposure at 1,380 ppm; and 2/8 following exposure at 1,040 ppm. These studies describe both lethal and nonlethal effects and are described in detail in Section 3.2.2.

No mortality was observed in guinea pigs (unspecified strain) exposed to HCl at 3,667 ppm for 5 min; however, at 4,333 ppm for 30 min or 667 ppm for 2 to 6 h, 100% mortality was observed (Machle et al. 1942).

3.1.2. Rats and Mice

Darmer et al. (1974) examined the acute toxicity of HCl vapor or aerosol in groups of 10 male Sprague-Dawley rats exposed to HCl at 410-30,000 ppm and groups of 10 male ICR mice exposed to HCl at 2,100-57,000 ppm for 5 or 30 min. HCl concentrations were monitored continuously during exposures, and chloride ion specific electrode analysis was utilized to determine actual concentrations. Particle size distribution was analyzed for aerosol generation. Animals were observed for 7 d post-exposure. Examination of animals dying during exposure revealed moderate to severe gross changes in the lungs and upper respiratory tract. Badly damaged nasal and tracheal epithelium, moderate to severe alveolar emphysema, atelectasis, and spotting of the lung were noted at necropsy. Survivors at the higher concentrations exhibited a clicking breathing noise, difficulty

TABLE 2-4 LC_{50} Values for Hydrogen Chloride Vapor and Aerosol in Rats and Mice (ppm)

	10-min LC_{50} Values		30-min LC_{50} Values	
Species	Vapor	Aerosol	Vapor	Aerosol
Rat	41,000	31,000	4,700	5,600
Mouse	13,700	11,200	2,600	2,100

Source: Darmer et al. 1974.

breathing, and bloody discharge from the nares. The authors conclude that there is no differential toxicity between the vapor and aerosol. Mice appear to be more sensitive than rats to the acute inhalation toxicity of HCl. LC_{50} values are presented in Table 2-4 (above).

Wohlslagel et al. (1976) also examined the acute toxicity of HCl in rats and mice (data are also reported in Vernot et al. 1977). Groups of 10 male CFE (Sprague-Dawley derived) rats and groups of 10 female CF-1 (ICR derived) mice were exposed to HCl vapor for 60 min. HCl concentrations were continuously monitored during exposures by specific ion analysis. Toxic signs observed during exposure included increased grooming and irritation of the eyes, mucous membranes, and exposed skin. By the end of the exposure period, rapid, shallow breathing and yellow-green fur discoloration were observed. Necropsy of animals that died during or after exposure revealed pulmonary congestion and intestinal hemorrhage in both species. Rats also showed thymic hemorrhages. Calculated LC_{50} values were 3,124 ppm for rats and 1,108 ppm for mice. As was also reported in the Darmer (1974) study, mice appear to be more sensitive than rats to the acute inhalation toxicity of HCl. Data are summarized in Table 2-5.

Higgins et al. (1972) also compared HCl toxicity in rats and mice. Groups of 10 Wistar rats and 15 ICR mice were exposed to various concentrations of HCl vapors for 5 min. HCl concentrations were monitored continuously during exposures via specific ion electrode analysis. Again, data suggest that mice are more sensitive to the lethal effects of HCl than are rats. Data are summarized in Table 2-6.

Buckley et al. (1984) exposed groups of 16-24 male Swiss-Webster mice (25-30 g) to HCl at 309 ppm (RD_{50}) 6 h/day (d) for 3 d. HCl concentrations were analyzed at least once per hour during exposures using infrared spectrometry. All mice were moribund or had died after the three exposures. Exfoliation, erosion, ulceration, and necrosis of the respiratory epithelium were observed.

TABLE 2-5 Mortality in Rats and Mice Exposed to Hydrogen Chloride for 60 Min

Rats		Mice	
Concentration (ppm)	Mortality	Concentration (ppm)	Mortality
1,813	0/10	557	2/10
2,585	2/10	985	3/10
3,274	6/10	1,387	6/10
3,941	8/10	1,902	8/10
4,455	10/10	2,476	10/10
LC_{50} (95% CI) = 3,124 ppm (2,829-3,450)		LC_{50} (95% CI) = 1,108 ppm (874-1,404)	

Abbreviation: CI, confidence interval.
Sources: Wohlslagel et al. 1976; Vernot et al. 1977.

In another study, Anderson and Alarie (1980) reported a 30-min LC_{50} value of 10,137 ppm for normal mice and a value of 1,095 ppm for trachea-cannulated mice.

3.1.3. Rabbits

No mortality was observed in rabbits (unspecified strain) exposed to HCl at 3,667 ppm for 5 min; however, at 4,333 ppm for 30 min or 667 ppm for 2 to 6 h, 100% mortality was observed (Machle et al. 1942).

3.2. Nonlethal Toxicity

3.2.1. Nonhuman Primates

Kaplan (1985) exposed juvenile male baboons (1 per concentration) to HCl at 190, 810, 2,780, 11,400, 16,570, or 17,290 ppm for 5 min. HCl exposure concentrations were continuously monitored using a "modified French standard test method." This method is based on continuous titration of the chloride ion with silver nitrate. The animals had been trained to perform an escape test. Escape was observed at 11,400 ppm and 17,290 ppm, and avoidance was observed at all other concentrations. The author

TABLE 2-6 Mortality in Rats and Mice Exposed to Hydrogen Chloride for 5 Min

Rats		Mice	
Concentration (ppm)	Mortality (%)	Concentration (ppm)	Mortality (%)
30,000	0	3,200	7
32,000	10	5,060	7
39,800	60	6,145	13
45,200	70	6,410	0
57,290	90	7,525	40
		8,065	13
		9,276	33
		13,655	40
		26,485	87
		30,000	87
LC_{50} (95% CI) = 40,989 ppm (34,803-48,272)		LC_{50} (95% CI) = 13,745 ppm (10,333-18,283)	

Abbreviation: CI, confidence interval.
Source: Higgins et al. 1972.

attributed these responses to irritation from the HCl exposure. No effects were noted at 190 ppm. Coughing and frothing at the mouth were observed in the animal exposed at 810 ppm, and shaking of the head, rubbing of the eyes, profuse salivation, and blinking were observed in animals exposed at higher concentrations. The baboons exposed to HCl at 16,570 ppm and 17,290 ppm died from bacterial infections several weeks after exposure. Necropsy indicated pneumonia, pulmonary edema, and tracheitis accompanied by epithelial erosion. In another study, Kaplan et al. (1988) exposed groups of three ketamine-anesthetized male baboons to HCl at target concentrations of 0, 500, 5,000, or 10,000 ppm for 15 min and observed for 3 mo. Exposures were accomplished in head boxes. HCl exposure concentrations were continuously monitored using a "modified French standard test method." Respiratory rates during exposure were increased approximately 30%, 50%, and 100% for the 500-, 5,000-, and 10,000-ppm groups, respectively. Arterial blood gas decreased 40% during the 15-min exposure at 5,000 ppm and 10,000 ppm, persisted for 10 min following exposure, and returned to baseline values by 3 d post-exposure. The increased respiratory rates were attributed to a compensatory response to the decrease in arterial oxygen, which in turn was attributed to upper airway broncho-constriction. No difference in pulmonary function or CO_2 challenge response tests was observed 3 d or 3 mo post-exposure when compared with baseline values. In a follow-up report (Kaplan et al. 1993a), no exposure-related differences in pulmonary function or CO_2 challenge response tests were reported 6 or 12 mo post-exposure, with the exceptions of respiratory rate at 5,000 ppm and 10,000 ppm, tidal volume at 5,000 ppm, and minute volume at 5,000 ppm and 10,000 ppm. Histopathologic examination was performed 12 mo post-exposure on one control animal, three animals in the 5,000-ppm group, and three animals in the 10,000-ppm group. One high-concentration animal exhibited pulmonary hemorrhage, edema, fibrosis, and bronchiolitis in the median right lung lobe. In another animal in the 10,000-ppm group, zonal atelectasis and focal multiple hemorrhages were observed in the right lung lobe. Focal, patchy hemorrhages were also observed in the three animals in the 5,000- ppm group as well as in the control group.

3.2.2. Guinea Pigs

Outbred English short-haired male guinea pigs (2-4 per group, weight 325-400 g) were exposed to HCl at 0, 107, 140, 162, or 586 ppm while exercising on a wheel (Malek and Alarie1989). Actual HCl concentrations were determined by sampling from a port just above the head of the running animal. Impingers containing 0.1 N sodium hydroxide were utilized. Animals had been exercising on the wheel for 10 min prior to the start of HCl exposure. Animals were exposed for 30 min or until incapacitation occurred. The 107-ppm group showed signs of mild irritation, while the other groups exhibited coughing and gasping prior to incapacitation. Summary data are presented in Table 2-7.

These data are in apparent conflict with other guinea pig data. For example, no deaths were observed after exposure to HCl at 500 ppm for 15 min, and three of six guinea pigs died after exposure at 4,200 ppm for 15 min (Kaplan et al. 1993b).

In another study, Burleigh-Flayer et al. (1985) exposed groups of male English smooth-haired guinea pigs (4-8 per group, weight 330-450 g) to HCl at 320, 680, 1,040, or 1,380 ppm for 30 min. HCl concentrations were measured colorimetrically. Sensory irritation was defined as decreased respiratory rate and a prolonged expiratory phase, and respiratory irritation

TABLE 2-7 Summary Data for Guinea Pigs Exposed to Hydrogen Chloride During Exercise

Concentration (ppm)	Number	% Incapacitated	Time to Incapacitation (min)	% Mortality
107 ± 26	3	0	NA	0
140 ± 5	3	100	16.5 ± 8.6	0
162 ± 0	2	100	1.3 ± 0.9	0
586 ± 51	4	100	0.65 ± 0.08	100[a]

[a]Time to death was 2.8 ± 0.8 min.
Abbreviation: NA, not applicable.
Source: Malek and Alarie 1989.

was defined as initial increased respiratory rate followed by a decrease due to a pause following each expiration. Sensory irritation was observed almost immediately in animals exposed at 680, 1,040, and 1,380 ppm, and appeared after 6 min of exposure in the 320-ppm group. Sensory irritation was replaced by pulmonary irritation in a concentration-related manner as sufficient HCl reached the deeper lung tissue. Time to onset of pulmonary irritation was approximately 18, 12, 7, and 3.5 min for the 320-, 680-, 1,040-, and 1,380-ppm groups, respectively. Corneal opacity was observed in one of four survivors from the 680-ppm group, five of five survivors from the 1,380-ppm group, and four of six from the 1,040-ppm group; no opacity was observed in the 320-ppm group. Histologic examination of the lungs was performed only on animals exposed at 1,040 ppm. Alveolitis accompanied by congestion and hemorrhage was observed 2 d post-exposure, and inflammation, hyperplasia, and mild bronchitis were observed 15 d post-exposure.

Kaplan et al. (1993b) exposed groups of six male English smooth-haired guinea pigs to HCl at 0, 500, or 4,200 ppm (nominal concentration) for 15 min. (Actual concentrations were 0, 520, or 3,940 ppm.) Exposure at 500 or 4,200 ppm caused a 20% decrease in respiratory frequency accompanied by a compensatory 2-fold increase in transpulmonary pressure. Exposure at 500 ppm had little effect on blood gases or arterial pH. Exposure at 4,200 ppm also had little effect on blood gases; however, a decrease in arterial pH was observed in exposed animals when compared with control animals. Necropsy performed 90 d post-exposure revealed lymphoid hyperplasia of the nasal mucosa and lungs and hemorrhage of the lungs in both control animals and animals exposed at 500 ppm. Exposure at 4,200 ppm resulted in the deaths of three guinea pigs at 2 min, 2.5 min, and 27 d post-exposure, respectively. Necropsy of those decedents showed pulmonary congestion, severe congestion of the nasal turbinates, severe tracheitis, and desquamation of bronchiolar epithelia. In the three guinea pigs of the 4,200-ppm exposure group necropsied 90 d post-exposure, cloudy corneas, focal hemorrhage, focal pneumonia, esophageal hyperkeratosis, atelectasis, and pulmonary lymphoid hyperplasia were observed.

3.2.3. Rats

Groups of three adult male Sprague-Dawley rats were exposed head-only to HCl at 200, 295, 784, 1,006, or 1,538 ppm for 30 min (Hartzell et al. 1985). Concentration-related decreases in respiratory frequency ranging from 35% to 67% were observed starting at approximately 2 min into the exposure period. Concentration-related decreases in minute-volume ranging from 30% to 69% were also observed.

Groups of eight male Fischer-344 rats were exposed to filtered air or HCl at 1,300 ppm for 30 min (Stavert et al. 1991). HCl concentrations were determined a minimum of three times per exposure using ion-specific electrodes. Each treatment had a nose-breathing group and a mouth-breathing group. Animals were sacrificed 24 h post-exposure. Nose-breathing rats exposed to HCl (actual concentration 1,293 ± 36 ppm) exhibited severe, necrotizing rhinitis, turbinate necrosis, thrombosis of nasal submucosa vessels, and pseudomembrane formation in the anterior portion of the nasal cavity. No effects were observed in the lungs of nose-breathers. Nasal cavities of mouth-breathers exposed to HCl (actual concentration 1,295 ± 25 ppm) were essentially unaffected; however, severe, ulcerative tracheitis accompanied by necrosis and luminal ulceration was observed. Polymorphonuclear leukocytes were observed in the submucosa between tracheal rings, in connective tissue around the trachea, and in alveoli surrounding terminal bronchioles in the mouth-breathers.

Kaplan et al. (1993b) exposed groups of six female Sprague-Dawley rats to HCl at 0 or 4,200 ppm (nominal concentration) for 15 min. (Actual concentrations were 0 or 3,890 ppm.) Exposure at 3,890 ppm caused a 40% decrease in respiratory frequency accompanied by a compensatory 1.2-fold increase in arterial pressure. Exposure at 3,890 ppm also caused a decrease in arterial pH, an increase in $PaCO_2$ values, and a transient decrease in PaO_2 values when compared with control animals. Necropsy performed 90 d post-exposure revealed microphthalmia in four rats and minimal focal atelectasis in one lobe of the lungs of two animals exposed at 3,890 ppm. Minimal focal atelectasis was also observed in two control animals.

3.2.4. Mice

Barrow et al. (1977) exposed groups of male Swiss-Webster mice (25-30 grams [g]) to HCl at concentrations ranging from 40 ppm to 943 ppm for 10 min. HCl concentrations were determined continuously using ion-specific electrodes. An RD_{50} of 309 ppm was defined.

In another study, Barrow et al. (1979) exposed groups of four male Swiss-Webster mice (25-30 g) to HCl at concentrations ranging from 20 ppm to 20,000 ppm for 10 min. Twenty-four hours after exposure, the mice were sacrificed by cervical dislocation. Precise individual group concentrations were not presented in the report. A decreased respiratory frequency, indicative of sensory irritation, was observed in animals exposed at >50 ppm. Deaths were observed in two of four mice exposed at 8,000 ppm and in four of four mice exposed at 19,300 ppm. Polymorphonuclear leukocyte infiltration of the conjunctiva was observed in animals exposed at 480 ppm, corneal necrosis was observed at 700 ppm, and ocular globe damage was observed at 3,000 ppm. Nasal epithelium ulceration was observed in the 120-ppm group, necrosis and damage to nasal bones in the 700-ppm group, and complete destruction of the nasal bones in the 7,000-ppm group.

Male Swiss-Webster mice were exposed to HCl at concentrations ranging from 17 ppm to 7,279 ppm for 10 min (Lucia et al. 1977). HCl concentrations were determined continuously using ion-specific electrodes. Animals were sacrificed 24 h post-exposure, and the upper respiratory tract was examined histologically. Small superficial ulcerations were observed in the respiratory epithelium at the junction with the squamous epithelium at 17 ppm. As HCl concentrations increased, the ulceration increased until it extended up the sides and into the nasal septum. The lower two-thirds of the upper respiratory tract was damaged at 723 ppm, and the entire mucosa was destroyed at 1,973 ppm. The squamous epithelium of the external nares was destroyed at 493 ppm, and the external support structures were destroyed at 1,088 ppm. Total destruction of mucosa and support structures as well as total destruction of the eyes was observed at 7,279 ppm.

Kaplan et al. (1993b) exposed groups of six male ICR mice to HCl at 0, 500, or 2,500 ppm (nominal concentration) for 15 min. (Actual concentrations were 0, 475, or 2,550 ppm.) Exposure to HCl at 475 ppm and 2,550 ppm caused 10% and 40% decreases in respiratory frequency, respectively, that were accompanied by compensatory increases in pressure. Four mice exposed at 475 ppm died within 90 d post-exposure, and all mice exposed at 2,550 ppm died by day 14 post-exposure. No histopathologic abnormalities were observed in controls or in mice exposed at 475 ppm. Moderate to severe lung congestion, necrosis of the tracheal mucosa, paranasal sinus exudate, and moderate lung edema were observed in animals exposed to HCl at 2,550 ppm. Results from this study are in apparent conflict with several other reported studies. For example, Darmer et al. (1974) reported a 5-min mouse LC_{50} of 13,700 ppm and a 30-min LC_{50} of 2,600 ppm, and Wholslagel et al. (1976) and Vernot et al. (1971) reported a 1-h LC_{50} of 1,108 ppm.

3.3. Developmental and Reproductive Toxicity

Female Wistar rats were exposed to HCl at 302 ppm for 1 h either 12 d prior to mating or on day 9 of gestation (Pavlova 1976). No information concerning test atmosphere concentration analysis was reported. One-third of the animals died, and dyspnea and cyanosis were observed. Congestion, edema, and hemorrhage were observed in the lungs of animals that died. Fetal mortality was higher ($p < 0.05$) in rats exposed during pregnancy and was possibly secondary to severe maternal effects. When female Wistar and mixed-strain rats were exposed to HCl at 302 ppm for 1 h prior to mating, 30% of the Wistar rats and 20% of the mixed-strain rats died from the exposure. In animals surviving 6 d, decreased blood oxygen saturation was observed as well as kidney, liver, and spleen damage. Treatment also altered the estrous cycle. In rats mated 12-16 d post-exposure and sacrificed on day 21 of gestation, increased fetal mortality, decreased fetal weight, and increased relative fetal lung weight were observed.

3.4. Genetic Toxicology

Genotoxicity results for HCl are equivocal. At a concentration of 25 µg/well, it was positive in a DNA repair assay in *E. Coli*, and it induced chromosomal nondisjunction in *D. melanogaster* at 100 ppm for 24 h. However, negative results were obtained in a Syrian hamster embryo cell transformation assay and in an adenovirus SA7 assay (NTIS 2000).

3.5. Carcinogenicity

Rats exposed to HCl at 10 ppm for 6 h/d, 5 d/wk for life developed increased incidences of tracheal and laryngeal hyperplasia compared with controls; however, no increase in the incidence of cancerous lesions was observed over controls (Sellakumar et al. 1985). Male Sprague-Dawley rats exposed at 10.2 ppm for 6 h/d, 5 d/wk for 382 exposures over a 588-d period did not show an increased incidence of nasal tumors (Albert et al. 1982).

3.6. Summary

HCl is a sensory and respiratory irritant and causes changes in the upper respiratory tract at relatively low concentrations and short exposure times. As concentrations and exposure times increase, effects progress to the lower respiratory tract and may involve pulmonary edema and histopathologic changes. There appears to be no differential toxicity between HCl vapor and aerosol, and mice appear to be more sensitive than rats to the effects of HCl. Exposure at 190 ppm for 5 min was the no-effect level in baboons, while those exposed at 16,570 ppm and 17,290 ppm exhibited pulmonary edema, pneumonia, and died from bacterial infections weeks after exposure. Mild irritation was observed in guinea pigs exposed at 107 ppm for 30 min, and guinea pigs exposed at concentrations ranging from 140 ppm to 586 ppm exhibited gasping prior to incapacitation and/or death. Exercising guinea pigs exposed to HCl at 320-1,380 ppm for 30 min exhibited a continuum of effects from sensory irritation to pulmonary irritation to death. Nose-breathing rats exposed to HCl exhibited severe nasal pathology, and similarly exposed mouth-breathing rats showed tracheal pathology. Several studies in mice also confirm the progression of effects from sensory irritation to lower respiratory tract involvement.

Fetal mortality was higher in rats exposed to HCl during pregnancy and 12-16 d prior to mating than it was in unexposed rats; however, no validation of exposure concentrations was provided. No data concerning the genotoxicity of HCl were located. In two lifetime studies, there was no increase in the incidence of cancerous lesions in rats exposed to HCl.

4. SPECIAL CONSIDERATIONS

4.1. Metabolism and Disposition

Information concerning the metabolism and disposition of hydrogen chloride (HCl) is sparse. HCl is not metabolized; however, hydrogen and chloride ions adsorbed in the respiratory tract may be distributed throughout the body (NRC 2000).

4.2. Mechanism of Toxicity

HCl is a respiratory irritant and is corrosive. When inhaled, it solubilizes in mucous present in the nasal passages, and when the scrubbing mechanism of the upper respiratory tract is saturated, it may enter the lower respiratory tract (NRC 1991). At a molecular level, HCl dissociates in water to form hydronium and chloride ions. The hydronium ion is a proton donor. It could catalyze cleavage of organic molecules and be involved in hydroxylation of carbonyl groups and polymerization and depolymerization of organic molecules (EPA 1994).

4.3. Structure-Activity Relationships

Although the AEGL values for HCl are based on empirical toxicity data, it is important to consider the relative toxicities of HCl and other structurally similar chemicals. The compounds most closely related to HCl are other hydrogen halides, HF and HBr. It might be anticipated that relationships exist in this chemical class between structure and respective toxicities in animals and humans. However, because of differences in size and electron configuration in the various halogen atoms, substantial differences exist with respect to their chemical and physical properties, which in turn are responsible for their toxicologic properties. That is particularly true in the case of acutely toxic effects resulting from inhalation exposure.

For example, HCl has a considerably higher ionization constant than HF, and is therefore classified as a stronger acid than HF. Consequently, higher concentrations of proton-donor hydronium ions are generated from HCl in aqueous solutions under the same conditions. The protons readily react with cells and tissues, resulting in HCl's irritant and corrosive properties. On the other hand, the fluoride ion from dissociated HF is a strong nucleophile or Lewis base that is highly reactive with various organic and inorganic electrophiles (biologically important substances), also resulting in irritation and tissue damage.

In addition to differences in chemical properties, difference in water solubility might be a significant factor in the acute inhalation toxicity of these substances. HF and HBr are characterized as infinitely and freely soluble in water, respectively, but the solubility of HCl, although still high, is lower (67 g/100 g of water at 30 C) (Budavari et al. 1989). Thus, it is likely that HF is more effectively scrubbed in the nasal cavity than HCl, resulting in less penetration to the lungs and less severe toxicity. The effectiveness of the scrubbing mechanism was demonstrated in a study that addressed the acute toxicities of HF, HCl, and HBr and the deposition (scrubbing) of those chemicals in the nasal passages. Stavert et al. (1991) exposed male Fischer-344 rats to each of the hydrogen halides at 1,300 ppm for 30 min and assessed damage to the respiratory tract 24 h after the exposure. The nasal cavity was divided into four regions and examined microscopically. For all three hydrogen halides, tissue injury was confined to the nasal cavity. Tissue injury in the nasal cavity was similar following exposures to HF and HCl and involved moderate to severe fibrinonecrotic rhinitis in nasal region 1 (most anterior region). For HF and HCl, the lesions extended into region 2, but regions 3 and 4 were essentially normal in appearance, as was the trachea. Nasal cavity lesions following exposure to HBr were limited to region 1 and were similar in extent to those produced by HF and HCl, showing that all three chemicals are well scrubbed. No lung or tracheal injury was evident for any of the chemicals, although accumulations of inflammatory cells and exudates in the trachea and lungs following exposure to HCl indicated that HCl may not be as well scrubbed in the nasal passages as HF and HBr. However, that possibility is modified by the authors' observation of lower minute-volumes in the HF- and HBr-exposed rats, so that greater amounts of HCl were inhaled. Morris and Smith (1982) also showed that at concentrations up to 226 ppm, >99.7% of inspired HF might be scrubbed in the upper respiratory tract of the rat.

In a series of experiments with HF and HCl that used guinea pigs and rabbits as the test species, Machle and coworkers (Machle and Kitzmiller 1935; Machle et al. 1934, 1942) concluded that the acute irritant effects of HF and HCl were similar, but the systemic effects of HF were more severe, presumably because chloride ion is a normal electrolyte in the body, and fluoride ion is not. However, the conclusions involving systemic effects followed repeated exposures.

Aside from lethality studies, no clear evidence is available to establish the relative toxicities of HCl and HF. At concentrations ranging from 100 ppm to 1,000 ppm for 30 min, Kusewitt et al. (1989) reported epithelial and submucosal necrosis, accumulation of inflammatory cells and exudates, and extravasation of erythrocytes in the nasal region of rats exposed to HF, HCl, or HBr. The severity of injury increased with increasing concentration, and the relative toxicities of the hydrogen halides were reported as HF > HCl > HBr. However, in a later study by the same authors, Stavert et al. (1991) reported no difference in the toxicities to the nasal regions or the lung in nose-breathing or mouth-breathing rats exposed to HF, HCl, or HBr at 1,300 ppm for 30 min.

At the high concentrations necessary to cause lethality during exposure durations of 5 min to 1 h, HF is approximately twice (1.8-2.2 times) as toxic to the rat as HCl (Table 2-8). The relationship is similar for the mouse within this time period (2.2-3.2 times); however, for respiratory irritants such as HCl, the mouse "may not be a good model for extrapolation to humans," because "mice appear to be much more susceptible to the lethal effects of HCl than other rodents or baboons....To some extent, this increased susceptibility may be due to less effective scrubbing of HCl in the upper respiratory tract" (NRC 1991). Quantitative data for HBr were limited to one study, but that study also showed that HF was more toxic than either HCl or HBr.

On the basis of empirical lethality (LC_{50}) data in rats, rabbits, and guinea pigs, the exposure time-LC_{50} relationship for HCl using the equation $C^n \times t = k$ results in an n value of 1. That is comparable to an n value of

TABLE 2-8 Relative Toxicities of HF, HCl, and HBr indicated by LC_{50} Values (ppm)

Species	Exposure Duration	HF	HCl	HBr	Reference
Rat	5 min	18,200	41,000		Higgins et al. 1972
Mouse		6,247	13,750		
Rat	30 min	2,042	4,700		Rosenholtz et al. 1963 (HF); MacEwan and Vernot 1972 (HCl)
Mouse			2,644		
Rat	1 h	1,395	3,124		Wohlslagel et al. 1976
Mouse		342	1,108		
Rat	1 h	1,278	2,350	2,858	MacEwan and Vernot 1972
Mouse		501	1,322	814	

Abbreviations: HBr, hydrogen bromide; HCl, hydrogen chloride; HF, hydrogen fluoride; LC_{50}, lethal concentration in 50% of subjects.

2 empirically derived from rat and mouse lethality data for HF. Hence, although HF is more toxic than HCl at the higher concentrations and shorter exposure durations, the rate of decrease in the LC_{50} threshold is less (i.e., less slope in the curve derived from $C^n \times t = k$) for HF than for HCl. As a result, the LC_{50} values, and therefore the lethal toxicities of HCl and HF, are comparable at 4 h and 8 h. This shift in relative lethal toxicity across time also is reflected in the AEGL-3 values developed for HCl and HF.

Considering the greater water solubility of HF compared with HCl, it is possible that the more effective scrubbing of HF in the nasal passages is responsible for the apparent decrease in the relative toxicities of HF and HCl at lower concentrations associated with longer exposure durations. Conversely, the greater toxicity of HF at higher concentrations associated with the shorter exposure durations might be due to saturation of the scrubbing mechanism and higher concentrations in the lower respiratory system.

4.4. Other Relevant Information

4.4.1. Species Variability

Differences in responses to HCl exposure have been observed between primates and rodents. Rodents exhibit sensory and respiratory irritation upon exposure to high concentrations of HCl. Dose-related decreases in respiratory frequency, indicative of a protective mechanism, are observed in rodents, while baboons inhaling HCl at 500, 5,000, or 10,000 ppm exhibited concentration-dependent increases in respiratory frequency, indicative of a compensatory response to hypoxia and a possible increase in the total dose delivered to the lung (NRC 1991).

Kaplan (1988) found that five of six mice died when exposed to HCl at 2,550 ppm for 15 min; however, baboons survived exposure at 10,000 ppm for 15 min. The LC_{50} values reported by Darmer et al. (1974), Wohlslagel et al. (1976), and Higgins et al. (1972) indicate that mice are approximately 3 times more sensitive than rats to the effects of HCl. That increased susceptibility might be due to less effective scrubbing of HCl in the upper respiratory tract of mice compared with rats (NRC 1991). Guinea pigs also appear to be more sensitive than rats; however, various guinea pig studies have provided conflicting results (Kaplan et al. 1993b).

Because most rodents are obligatory nose-breathers, whereas humans may be mouth-breathers, especially during exercise, Stavert et al. (1991)

studied the effects of inhalation of HCl via the nose and mouth in rats. HCl was delivered directly to the trachea by cannulation; concentrations that produced effects confined to the nasal passages in nose-breathing rats resulted in serious lower respiratory tract effects and/or deaths in orally cannulated rats. These results indicate that the site of injury and resultant toxicologic effects differ with mouth- and nose-breathing, the former mode resulting in more severe responses under similar exposure situations. Thus, species that breathe through their mouth (humans) may be more sensitive to the effects of HCl than those that are obligate nasal breathers (rodents).

4.4.2. Unique Physicochemical Properties

HCl gas is hygroscopic and dissolves in water to produce hydrochloric acid. Concentrated hydrochloric acid is 37% HCl. When HCl gas absorbs moisture, it is highly reactive with metals and releases hydrogen gas (EPA 1994).

4.4.3. Concurrent Exposure Issues

Gases such as carbon monoxide, carbon dioxide, hydrogen cyanide, nitrogen dioxide, and nitrogen are often present with HCl during fires. The temperature, moisture content, and particular mix of gases can influence toxicity. For example, carbon monoxide has been shown to weaken the irritant effects of HCl (Sakurai 1989). However, Higgins et al. (1972) found no effect on mortality of rats and mice exposed to HCl alone or in combination with carbon monoxide.

Wohlslagel et al. (1976) examined the hazard of simultaneous acute inhalation exposure to HCl, HF, and alumina dust, which are components of solid rocket motor exhaust. Sixty-minute LC_{50} values were determined for both rats and mice. Exposures were to HCl alone, HF alone, and combination exposures to HCl and HF. Data suggested that in both species exposure to both gases simultaneously resulted in physiologically additive mortality incidences. No synergism, potentiation, or antagonism were observed. Gross and histopathologic examinations of rats and mice exposed to combinations of the corrosive gases showed no sites of damage additional to those observed in animals exposed to a single gas. The addition of alumina dust to atmospheres containing HCl and HF vapors did not increase or decrease mortality in either species.

HCl in smoke generated by flaming thermodegradation polyvinyl chloride (30-min LC_{50} of 2,141 ppm) was slightly more lethal to rats than HCl in smoke generated by nonflaming thermodegradation (30-min LC_{50} of 2,924 ppm), which was slightly more lethal than pure HCl (30-min LC_{50} of 3,817 ppm) (Hartzell et al. 1987).

4.4.4. Subchronic Exposure

Groups of 31 male and 21 female rats or 31 male and 21 female mice were exposed to HCl at 0, 10, 20, or 50 ppm for 6 h/d, 5 d/wk for 90 d (Toxigenics 1984). After the fourth exposure, 15 males and 10 females per group per species were sacrificed. Decreased body weight was observed in both genders of both species exposed at 50 ppm for 4 d. Hematology and urinalysis were unremarkable for those animals. Rats exhibited minimal to mild rhinitis after exposure to the three HCl concentrations for 4-90 d. The rhinitis was concentration- and duration-related and occurred in the anterior portion of the nasal cavity. After 90 d, mice in the 50-ppm group developed varying degrees of cheilitis with accumulations of hemosiderin-laden macrophages in the perioral tissues. In addition, at all three concentrations, mice developed signs of minor, reversible degeneration (eosinophilic globules) in the epithelial cells lining the nasal turbinates.

Machle et al. (1942) exposed three rabbits, three guinea pigs, and one monkey to HCl at 34 ppm for 6 h/d, 5 d/wk for 4 wk. No histopathologic effects were noted at necropsy. Clinical signs during or after exposure and other experimental details were not reported.

Mice exposed to HCl at 310 ppm for 6 h/d for 5 d exhibited necrosis, exfoliation, erosion, and ulceration of the nasal respiratory epithelium; however, no histopathologic effects were noted in the lung (Buckley et al. 1984).

5. RATIONALE AND PROPOSED AEGL-1

AEGL-1 is the airborne concentration (expressed as parts per million or milligrams per cubic meter [ppm or mg/m^3]) of a substance above which it is predicted that the general population, including susceptible individuals, could experience notable discomfort, irritation, or certain asymptomatic, nonsensory effects. However, the effects are not disabling and are transient and reversible upon cessation of exposure.

5.1. Summary of Human Data Relevant to AEGL-1

Five male and five female exercising adult asthmatic subjects (ages 18-25 y) were exposed to filtered air or HCl at 0.8 ppm or 1.8 ppm for 45 min (Stevens et al. 1992). No treatment-related effects were observed.

5.2. Summary of Animal Data Relevant to AEGL-1

No animal data consistent with effects defined by AEGL-1 were available.

5.3. Derivation of AEGL-1

Because appropriate human data exist for exposure to HCl, they were used to identify AEGL-1 values. Exposure to HCl at 1.8 ppm for 45 min resulted in a no-observed-adverse-effect level (NOAEL) in 10 exercising young adult asthmatic subjects (Stevens et al. 1992). Because exercise will increase HCl uptake and exacerbate irritation, those asthmatic subjects are considered a sensitive subpopulation. Therefore, because the test subjects were a sensitive subpopulation and the end point was essentially a no-effect level, no uncertainty factory (UF) was applied to account for sensitive human subpopulations. Adequate human data were available, so no UF was applied for animal to human extrapolation. The no-effect level was held constant across the 10- and 30-min and 1-, 4-, and 8-h exposure time points. That approach was considered appropriate because mild irritant effects generally do not vary greatly over time, and the end point of a no-effect level in a sensitive population is inherently conservative. The values for AEGL-1 are given in Table 2-9. Figure 2-1 is a plot of the derived AEGLs and the human and animal data on HCl.

6. RATIONALE AND PROPOSED AEGL-2

AEGL 2 is the airborne concentration (expressed as ppm or mg/m^3) of a substance above which it is predicted that the general population, including susceptible individuals, could experience irreversible or serious, long lasting adverse health effects or an impaired ability to escape.

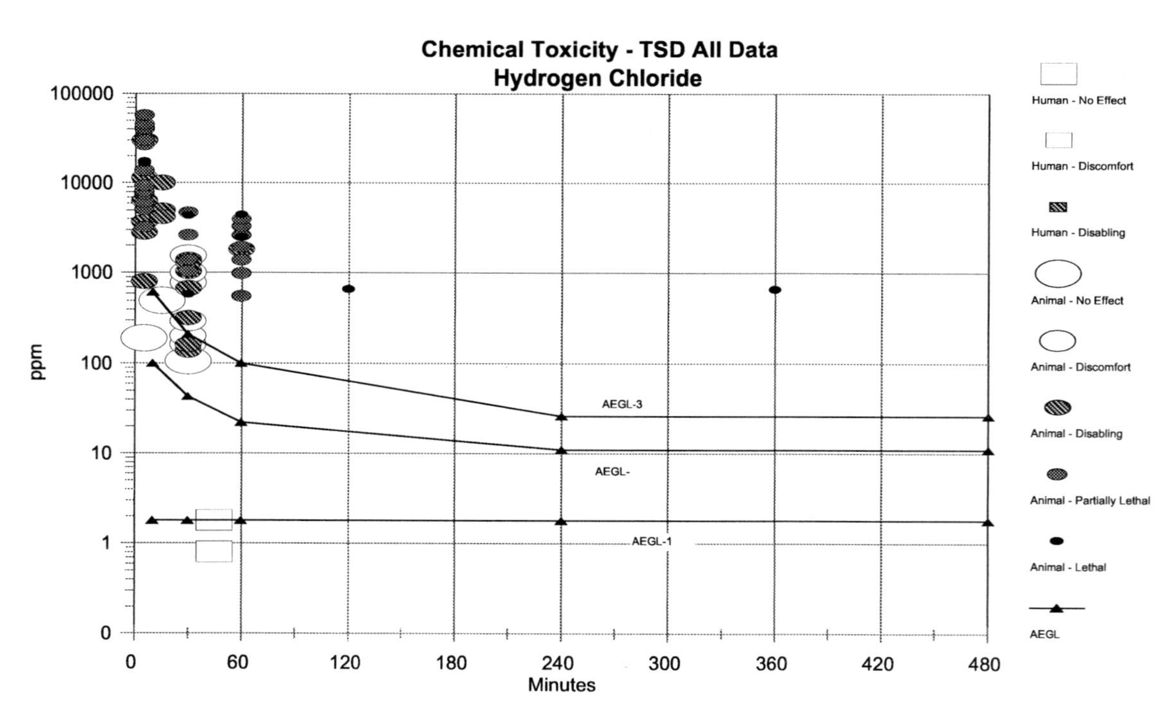

FIGURE 2-1 Toxicity data and AEGL values for hydrogen chloride. Toxicity data include both human and animal studies.

TABLE 2-9 AEGL-1 Values for Hydrogen Chloride (ppm [mg/m^3])

10 min	30 min	1 h	4 h	8 h
1.8 (2.7)	1.8 (2.7)	1.8 (2.7)	1.8 (2.7)	1.8 (2.7)

6.1. Summary of Human Data Relevant to AEGL-2

Human data consistent with effects defined by AEGL-2 from exposure to HCl are not appropriate for use in derivation because the descriptions of concentration, exposure time, and effects are subjective.

6.2. Summary of Animal Data Relevant to AEGL-2

Studies in baboons involving escape performance tests (Kaplan 1987) and pulmonary function (Kaplan et al. 1988, 1993a) resulted in effects consistent with those defined by AEGL-2. However, the baboon study descriptions suggest that individual baboons may have been used for more than one exposure. Responses of both guinea pigs (Malek and Alarie 1989; Burleigh-Flayer et al. 1985) and rats (Stavert et al. 1991) also are consistent with effects defined by AEGL-2. Other studies, although designed fairly well, produced effects more severe than that defined by AEGL-2 (Darmer et al. 1974; Buckley et al. 1984; Barrow et al. 1979; Hartzell et al. 1985).

6.3. Derivation of AEGL-2

The AEGL-2 for the 30-min and 1-, 4-, and 8-h time points was based on severe nasal or pulmonary histopathology in rats exposed to HCl at 1,300 ppm for 30 min (Stavert et al. 1991). A modifying factor (MF) of 3 was applied to account for the relatively sparse database describing effects defined by AEGL-2. The AEGL-2 values were further adjusted by a total UF of 10—3 for intraspecies variability, supported by the steep concentration-response curve, which implies little individual variability; and 3 for interspecies variability. Using the default value of 10 for interspecies variability would bring the total adjustment to 100 instead of 30. That would generate AEGL-2 values that are not supported by data on exercising asthmatic subjects, an especially sensitive subpopulation. Exercise increases HCl uptake and exacerbates irritation; no effects were noted in exercising young adult asthmatic subjects exposed to HCl at 1.8 ppm for 45 min (Stevens et al. 1992). Using a total UF of 30 would yield 4- and 8-h values of 3.6 ppm (instead of 11 ppm). The prediction that humans would be disabled by exposure for 4 h or 8 h to 3.6 ppm cannot be supported when exercising asthmatic subjects exposed to one-half that concentration for 45 min exhibited no effects. The shorter time points would yield values 4 to 7 times the 1.8-ppm value; however, confidence in the time-scaling for HCl is good for times up to 100 min, because the value of n was derived from a regression analysis of rat and mouse mortality data with exposure durations ranging from 1 min to 100 min. The 30-min value of 43 ppm derived with a total UF of 10 is reasonable in light of the fact that baboons exposed at 500 ppm for 15 min experienced only a slightly increased respiratory rate. Therefore, a total UF of 10, accompanied by the MF of 3, is most consistent with the database. Thus, the total factor is 30. Time-scaling for the 1-h AEGL exposure period used the $C^n \times t = k$ relationship, where $n = 1$ based on regression analysis of combined rat and mouse LC_{50} data (1 min to 100 min) as reported by ten Berge et al. (1986). The 4- and 8-h AEGL-2 values were derived by applying an MF of 2 to the 1-h AEGL-2 value, because time-scaling would yield a 4-h AEGL-2 of 5.4 ppm and an 8-h AEGL-2 of 2.7 ppm, close to the 1.8 ppm tolerated by exercising asthmatic subjects without observed adverse health effects. Repeated-exposure rat data suggest that the 4- and 8-h values of 11 ppm are protective. Rats exposed to HCl at 10 ppm for 6 h/d, 5 d/wk for life exhibited only tracheal and laryngeal hyperplasia, and rats exposed to HCl at 50 ppm for 6 h/d, 5 d/wk for 90 d exhibited only mild rhinitis.

The 10-min AEGL-2 was derived by dividing the mouse RD_{50} of 309 ppm by a factor of 3 to obtain a concentration causing irritation (Barrow 1977). It has been determined that human response to sensory irritants can be predicted on the basis of the mouse RD_{50}. For example, Schaper (1993) has validated the correlation of $0.03 \times RD_{50} = TLV$ as a value that will prevent sensory irritation in humans. The 0.03 represents the half-way point between 0.1 and 0.01 on a logarithmic scale, and Alarie (1981) has shown that the RD_{50} multiplied by 0.1 corresponds to "some sensory irritation," whereas the RD_{50} value itself is considered "intolerable to humans." Thus, it is reasonable that one-third of the RD_{50}, a value half-way between 0.1 and 1 on a logarithmic scale, might cause significant irritation to humans. Furthermore, one-third of the mouse RD_{50} for HCl corresponds to an approximate decrease in respiratory rate of 30%, and decreases in the range of 20-50% correspond to moderate irritation (ASTM 1991). The values for AEGL-2 are given in Table 2-10.

TABLE 2-10 AEGL-2 Values for Hydrogen Chloride (ppm [mg/m^3])

10 min	30 min	1 h	4 h	8 h
100 (156)	43 (65)	22 (33)	11 (17)	11 (17)

The 10-min value is supported by the baboon studies, where no effects were noted in a baboon exposed at 190 ppm for 5 min (Kaplan 1987), and no exposure-related effects on tidal volume or PaO_2 were observed in anesthetized baboons exposed at 500 ppm for 15 min (Kaplan et al. 1988). Also, when the rat data of Stavert et al. (1991) are extrapolated back to 10 min, a value of 130 ppm is obtained, suggesting that the proposed 10-min value is protective.

7. RATIONALE AND PROPOSED AEGL-3

AEGL-3 is the airborne concentration (expressed as ppm or mg/m^3) of a substance above which it is predicted that the general population, including susceptible individuals, could experience life-threatening health effects or death.

7.1. Summary of Human Data Relevant to AEGL-3

No data concerning human lethality from HCl exposure were located in the available literature.

7.2. Summary of Animal Data Relevant to AEGL-3

Baboon exposure studies involving an escape performance test and pulmonary function tests resulted in effects consistent with those defined by AEGL-3 (Kaplan 1987; Kaplan et al. 1988). Other well-designed studies identified disabling effects and lethality in guinea pigs (Malek and Alarie 1989; Burleigh-Flayer et al. 1985), rats, and mice (Darmer et al. 1974; Wohlslagel et al. 1976; Vernot et al. 1977; Barrow et al. 1979; Buckley et al. 1984; Hartzell et al. 1985).

7.3. Derivation of AEGL-3

The AEGL-3 was based on a 1-h rat LC_{50} study (Wohlslagel et al. 1976; Vernot et al. 1977). One-third of the 1-h LC_{50} value of 3,124 ppm was used as an estimated concentration causing no deaths. That estimate is inherently conservative (no deaths observed in the same study at 1,813 ppm). A total UF of 10 will be applied—3 for intraspecies variation, because the steep concentration-response curve implies limited individual variability; and 3 to protect susceptible individuals. Using a full value of 10 for interspecies variability (total UF of 30) would yield AEGL-3 values that are inconsistent with the overall data set.

A number of factors argue for the use of a UF of 10 instead of 30: (1) the steep concentration-response curve for lethality observed in the Wohlslagel et al. (1976) study in which the estimated LC_0 (one-third of the LC_{50} of 3,124 ppm) is lower than the experimental LC_0 of 1,813 ppm. The LC_0 selection is conservative, and the steep concentration-response curve argues for little interindividual variability; (2) AEGL-3 values generated from a total UF of 30 would be close (within a factor of 2) to the AEGL-2 values generated from data on exercising asthmatic subjects; (3) Sellakumar et al. (1985) exposed rats to HCl at 10 ppm for 6 h/d, 5 d/wk for life and only observed increased trachael and laryngeal hyperplasia. The estimated 6-h AEGL-3 using an intraspecies UF of 3 is 17 ppm, close to the concentration inhaled in the lifetime study in which only mild effects were induced; and (4) rats exposed to HCl at 50 ppm for 6 h/d, 5 d/wk for 90 d (Toxigenics 1984) exhibited mild rhinitis. This level is already twice the AEGL-3 value, which is intended to protect against death.

Thus, the total UF was set at 10. It was then time-scaled to the specified 10- and 30-min and 4-h AEGL exposure periods using the $C^n \times t = k$ relationship, where $n = 1$ based on regression analysis of combined rat and mouse LC_{50} data (1 min to 100 min) as reported by ten Berge et al. (1986). The 4-h AEGL-3 also was adopted as the 8-h AEGL-3 because of the uncertainty of time-scaling to 8 h with an n value derived from exposure durations of up to 100 min. The values for AEGL-3 are given in Table 2-11.

The 5-min rat LC_0 of 30,000 ppm (Higgins et al. 1972) supports the 10-min AEGL-3 value. Extrapolating that value across time ($n = 1$) to 10 min and applying a UF of 10 yields a value of 1,500 ppm, suggesting that the proposed AEGL-3 value is protective. Also, if the 5-min rat LC_{50} of 41,000 ppm for HCl vapor (Darmer et al. 1974) is divided by 3 to estimate a no-effect level for death, extrapolated to 10 min, and a UF of 10 is applied, a supporting value of 683 ppm is obtained.

TABLE 2-11 AEGL-3 Values for Hydrogen Chloride (ppm [mg/m^3])

10 min	30 min	1 h	4 h	8 h
620 (937)	210 (313)	100 (155)	26 (39)	26 (39)

8. SUMMARY OF PROPOSED AEGLS

8.1. AEGL Values and Toxicity End Points

The derived AEGLs for various levels of effects and durations of exposure are summarized in Table 2-12. A NOAEL for sensory irritation in exercising asthmatic subjects was used for AEGL-1. Severe nasal and pulmonary effects in rats and a modification of the mouse RD_{50} were used for AEGL-2. An estimated no-effect level for death in rats was used for AEGL-3.

8.2. Other Exposure Criteria

Standards set by other organizations appear in Table 2-13.

8.3. Data Quality and Research Needs

Human data are limited to one study showing no significant effects in asthmatic subjects and to dated anecdotal information. Furthermore, the

TABLE 2-12 Summary of AEGL Values for Hydrogen Chloride (ppm [mg/m^3])

Classification	10 min	30 min	1 h	4 h	8 h
AEGL-1 (Nondisabling)	1.8 (2.7)	1.8 (2.7)	1.8 (2.7)	1.8 (2.7)	1.8 (2.7)
AEGL-2 (Disabling)	100 (156)	43 (65)	22 (33)	11 (17)	11 (17)
AEGL-3 (Lethal)	620 (937)	210 (313)	100 (155)	26 (39)	26 (39)

TABLE 2-13 Extant Standards and Guidelines for Hydrogen Chloride (ppm)

	Exposure Duration				
Guideline	10 min	30 min	1 h	4 h	8 h
AEGL-1	1.8	1.8	1.8	1.8	1.8
AEGL-2	100	43	22	11	11
AEGL-3	620	210	100	26	26
ERPG-1[a]	3				
ERPG-2[a]	20				
ERPG-3[a]	150				
NIOSH IDLH[b]	50				
NIOSH REL[c]					5 (ceiling)
OSHA PEL-TWA[d]					5 (ceiling)
ACGIH TLV-STEL[e]	5				
NRC SPEGL[f]					1
NRC EEGL[g]			20		
NRC SMAC[h]	5				
German MAK[i]					5
Dutch MAC[j]					5

[a]ERPG (emergency response planning guidelines) of the American Industrial Hygiene Association (AIHA 2001). ERPG-1 is the maximum airborne concentration below which it is believed nearly all individuals could be exposed for up to 1 h without experiencing symptoms other than mild, transient adverse health effects or without perceiving a clearly defined objectionable odor. The ERPG-1 for HCl is based on objectionable odor. The ERPG-2 is the maximum airborne concentration below which it is believed nearly all individuals could be exposed for up to 1 h without experiencing or developing irreversible or other serious health effects or symptoms that could impair an individual's ability to take protection action. The ERPG-2 for HCl is based on animal studies suggesting serious eye and respiratory irritation above 20 ppm and below 100 ppm. The ERPG-3 is the maximum airborne concentration below which it is believed nearly all individuals could be exposed for up to 1 h without experiencing or developing life-threatening health effects. The ERPG-3 for HCl is based on animal data suggesting that concentrations exceeding 150 ppm for 1 h may produce severe, possibly life-threatening health effects, such as pulmonary edema, in a heterogeneous population.
[b]IDLH (immediately dangerous to life and health standard of the National Institute of Occupational Safety and Health) (NIOSH 1994). The IDLH represents the maximum concentration from which one could escape within 30 min without any

escape-impairing symptoms or irreversible health effects. The IDLH for HCl is based on acute inhalation toxicity data in humans.
[c]NIOSH REL-STEL (recommended exposure limits-short-term exposure limit) (NIOSH 1997). The REL-STEL is defined analogous to the ACGIH TLV-TWA.
[d]OSHA PEL-TWA (permissible exposure limit-time-weighted average of the Occupational Safety and Health Administration) (OSHA 1997). The PEL-TWA is defined analogous to the ACGIH TLV-TWA but is for exposures of no more than 10 h/d, 40 h/wk.
[e]ACGIH TLV-STEL (Threshold Limit Value-short-term exposure limit of the American Conference of Governmental Industrial Hygienists) (ACGIH 2000). The TLV-STEL for HCl is based on corrosion and irritation.
[f]SPEGL (short-term public emergency guidance level) (NRC 1991).
[g]EEGL (emergency exposure guidance level) (NRC 1985). The EEGL is the concentration of contaminants that can cause discomfort or other evidence of irritation or intoxication in or around the workplace, but avoids death, other severe acute effects, and long-term or chronic injury.
[h]SMAC (spacecraft maximum allowable concentration of NASA) (NRC 2000).
[i]MAK (Maximale Argeitsplatzkonzentration [Maximum Workplace Concentration]) (Deutsche Forschungsgemeinschaft [German Research Association] 2000). The MAK is defined analogous to the ACGIH TLV-TWA.
[j]MAC (Maximaal Aanvaaarde Concentratie [Maximal Accepted Concentration]) (SDU Uitgevers [under the auspices of the Ministry of Social Affairs and Employment], The Hague, The Netherlands 2000). The MAC is defined analogous to the ACGIH TLV-TWA.

involvement of RADS in HCl toxicity is unclear. Many more data are available for animal exposures; however, many of those studies used compromised animals or very small experimental groups, resulting in limited data for many species but no in- depth database for a given species. Also, some studies involve very short exposures to high concentrations of HCl. Thus, confidence in the AEGL values is at best moderate.

9. REFERENCES

AIHA (American Industrial Hygiene Association). 2001. Emergency Response Planning Guidelines. Hydrogen Chloride. Fairfax, VA: AIHA.

Alarie, Y. 1981. Dose-response analysis in animal studies: Prediction of human responses. Environ. Health Perspect. 42:9-13.

Albert, R.E., A.R. Sellakumar, S. Laskin, M. Kuschner, N. Nelson, and C.A. Snyder. 1982. Gaseous formaldehyde and hydrogen chloride induction of nasal cancer in the rat. JNCI 68:597-603.

Amoore, J.E., and E. Hautala. 1983. Odor as an aid to chemical safety: Odor thresholds compared with threshold limit values and volatiles for 214 industrial chemicals in air and water dilution. J. Appl. Toxicol. 3: 272- 290.

Anderson, R.C., and Y. Alarie. 1980. Acute lethal effects of polyvinylchloride thermal decomposition products in normal and cannulated mice. The Toxicologist A3.

ASTM (American Society for Testing and Materials). 1991. Method E981. Pp. 610-619 in Standard Test Method For Estimating Sensory Irritancy of Airborne Chemicals, Volume 11.04. Philadelphia, PA: ASTM.

Barrow, C.S., H. Lucia, and Y.C. Alarie. 1979. A comparison of the acute inhalation toxicity of hydrogen chloride versus the thermal decomposition products of polyvinylchloride. J. Combust. Toxicol. 6:3-12.

Barrow, C.S., Y. Alarie, M. Warrick, and M.F. Stock. 1977. Comparison of the sensory irritation response in mice to chlorine and hydrogen chloride. Arch. Environ. Health 32:68-76.

Beaumont, J.J., J. Leveton, K. Knox, T. Bloom, T. McQuiston, M. Young, R. Goldsmith, N.K. Steenland, D.P. Brown, and W.E. Halperin. 1987. Lung cancer mortality in workers exposed to sulfuric acid mist and other acid mists. JNCI 79:911-921.

Bernstein, J.A. 1993. Reactive Airways Dysfunction Syndrome (RADS). DPICtions 12:1-3.

Bond, G.G., R.R. Cook, P.C. Wright, and G.H. Flores. 1983. A case-control study of brain tumor mortality at a Texas chemical plant. J. Occup. Med. 25:377-386.

Bond, G.G., R.J. Shellenburger, G.H. Flores, R.R. Cook, and W.A. Fiskbeck. 1985. A case-control study of renal cancer mortality at a Texas chemical plant. Am. J. Ind. Med. 7:123-139.

Bond, G.G., G.H. Flores, R.J. Shellenburger, J.B. Cartmill, W.A. Fiskbeck, and R.R. Cook. 1986. Nested case-control study of lung cancer among chemical workers. Am. J. Epidemiol. 124:53-66.

Bond, G.G., G.H. Flores, B.A. Stafford, and G.W. Olsen. 1991. Lung cancer and hydrogen chloride exposure: Results from a nested case-control study of chemical workers. J. Occup. Med. 33:958-961.

Boulet, L-P. 1988. Increases in airway responsiveness following acute exposure to respiratory irritants: Reactive airway dysfunction syndrome or occupational asthma? Chest 94:476-481.

Brooks, S.M., M.A. Weiss, and I.L. Bernstein. 1985. Reactive airways dysfunction syndrome (RADS): Persistent asthma syndrome after high level irritant exposures. Chest 88:376-384.

Buckley, L.A., X.Z. Jiang, R.A. James, K.T. Morgan, and C.S. Barrow. 1984. Respiratory tract lesions induced by sensory irritants at the RD50 concentration. Toxicol. Appl. Pharmacol. 74:417-429.

Burleigh-Flayer, H., K.L. Wong, and Y. Alarie. 1985. Evaluation of the pulmonary effects of HCl using CO2 challenges in guinea pigs. Fundam. Appl. Toxicol. 5:978-985.

Darmer, K.L., E.R. Kinkead, and L.C. DiPasquale. 1974. Acute toxicity in rats and mice exposed to hydrogen chloride gas and aerosols. Am. Ind. Hyg. Assoc. J. 35:623-631.

DFG (Deutsche Forschungsgemeinschaft). 2000. List of MAK and BAT Values, Commission for the Investigation of Health Hazards of Chemical Compounds in the Work Area, Report No. 35 [in German]. Weinheim, Federal Republic of Germany: Wiley VCH.

Elkins, H.B. 1959. The Chemistry of Industrial Toxicology, 2nd Ed. New York, NY: John Wiley & Sons. Pp. 79-80.

EPA (U.S. Environmental Protection Agency) 1994. Health Assessment Document for Chlorine and Hydrogen Chloride. External Review Draft. ECAO-R-065. Environmental Criteria and Assessment Office, Research Triangle Park, NC.

EPA (U.S. Environmental Protection Agency). 1995. Hydrogen Chloride. Integrated Risk Information System (IRIS) [Online]. Available: http://www.epa.gov/iris/subst/0396.htm.

Hartzell, G.E., H.W. Stacy, W.G. Swiztzer, D.N. Priest, and S.C. Packham. 1985. Modeling of toxicological effects of fire gases: IVV. Intoxication of rats by carbon monoxide in the presence of an irritant. J. Fire Sci. 3:263-279.

Hartzell, G.E., A.F. Grand, and W.G. Swiztzer. 1987. Modeling of toxicological effects of fire gases: VI. Further studies on the toxicity of smoke containing hydrogen chloride. J. Fire Sci. 5:368-391.

Henderson, Y., and H. W. Hagard. 1943. Noxious Gases. New York: Reinhold Publishing Corp. Pp. 126.

Higgins, E.A., V. Fiorca, A.A. Thomas, and H.V. Davis. 1972. Acute toxicity of brief exposures to HF, HCL, NO2, and HCN with and without CO. Fire Tech. 8:120-130.

HSDB (Hazardous Substances Data Bank). 1996. Hydrogen chloride [Online]. Available: http://toxnet.nlm.nih.gov/cgi-bin/sis/search/f?./temp/~LRWhf3:1 [July 22, 1996].

IARC (International Agency for Research on Cancer). 1992. IARC Monographs on the Evaluation of Carcinogenic Risks to Humans, Vol. 54: Occupational Exposures to Mists and Vapors from Strong Inorganic Acids; and Other Industrial Chemicals. Lyon, France: IARC. Pp. 189-211.

Kaplan, H.L., et al. 1985. Effects of combustion gasses on escape performance in the baboon and rat. J. Fire Sci. 3:228-244.

Kaplan, H.L. 1987. Effects of irritant gases on the avoidance/escape performance and respiratory response of the baboon. Toxicology 47:165-179.

Kaplan, H.L., A. Anzeuto, W.G. Switzer, and R.K. Hinderer. 1988. Effects of hydrogen chloride on respiratory response and pulmonary function of the baboon. J. Toxicol. Environ. Health 23:473-493.

Kaplan, H.L., W.G. Switzer, R.K. Hinderer, and A. Anzeuto. 1993a. A study on the acute and long-term effects of hydrogen chloride on respiratory response and pulmonary function and morphology in the baboon. J. Fire Sci. 11:459-484.

Kaplan, H.L., W.G. Switzer, R.K. Hinderer, and A. Anzeuto. 1993b. Studies of the effects of hydrogen chloride and polyvinyl chloride (PVE) smoke in rodents. J. Fire Sci. 11:512-552.

Kusewitt, D.F., D.M. Stavert, G. Ripple, T. Mundie, and B.E. Lehnert. 1989. Relative acute toxicities in the respiratory tract of inhaled hydrogen fluoride, hydrogen bromide, and hydrogen chloride. Toxicologist 9:36.

Leonardos, G., Kendall, and N.J. Barnard. 1969. Odor threshold determinations of 53 odorant chemicals. J. Air Pollut. Control Assoc. 19:91-95.

Lucia, H.L., C.S. Barrow, M.F. Stock, and Y. Alarie. 1977. A semi-quantitative method for assessing anatomic damage sustained by the upper respiratory tract of the laboratory mouse, Mus musculis. J. Combust. Toxicol. 4:472-486.

Machle, W., K.V. Kitzmiller, E.W. Scott, and J.F. Treon. 1942. The effect of inhalation of hydrogen chloride. J. Ind. Hyg. Toxicol. 22:222-225.

Malek, D.E., and Y. Alarie. 1989. Ergometer within a whole-body plethysmograph to evaluate performance of guinea pigs under toxic atmospheres. Toxicol. Appl. Pharmacol. 101:340-355.

Nemery, B. 1996. Late consequences of accidental exposure to inhaled irritants: RADS and the Bhopal disaster. Eur. Respir. J. 9:1973-1976.

NIOSH (National Institute for Occupational Safety and Health). 1994. Documentation for Immediately Dangerous to Life or Health Concentrations (IDLHs). U.S. Department of Health and Human Services, National Institute for Occupational Safety and Health, Cincinnati, OH.

NRC (National Research Council). 1987. Emergency and Continuous Exposure Guidance Levels for Selected Airborne Contaminants, Vol. 7. Washington DC: National Academy Press.

NRC (National Research Council). 1991. Permissible Exposure Levels and Emergency Exposure Guidance Levels for Selected Airborne Contaminants. Washington, DC: National Academy Press. Pp. 37-52.

NRC (National Research Council). 2000. Spacecraft Maximum Allowable Concentrations for Selected Airborne Contaminants, Vol. 4. Washington DC: National Academy Press.

NTIS (National Technical Information Service). 2000. Hydrogen Chloride. Registry of Toxic Effects of Chemical Substances (RTECS) [Online]. Available: http://www.ntis.gov/search/product.asp?ABBR=SUB5363&starDB=GRAHIST [October 1, 2000].

OSHA (Occupational Safety and Health Administration). 1999. CFR 29 Part 1910. Occupational Safety and Health Standards. Air Contaminants. U.S. Department of Labor, Washington, DC.

Pavlova, T.E. 1976. Disturbance of development of the progeny of rats exposed to hydrogen chloride. Bull. Exp. Biol. Med. 82:1078-1081.

Promisloff, R.A., G.S. Lenchner, and A.V. Cichelli. 1990. Reactive airway dysfunction syndrome in three police officers following a roadside chemical spill. Chest 98:928-929.

Sakurai, T. 1989. Toxic gas tests with several pure and mixed gases using mice. J. Fire. Sci. 7:22-77.

Schaper, M. 1993. Development of a database for sensory irritants and its use in establishing occupational exposure limits. Am. Ind. Hyg. Assoc. J. 54:488-544.

Ministry of Social Affairs and Employment (SDU Uitgevers). 2000. National MAC (Maximum Allowable Concentration) List, 2000. Ministry of Social Affairs and Employment, The Hague, The Netherlands.

Sellakumar, A.R., C.A. Snyder, J.J. Solomon, and R.E. Albert. 1985. Carcinogenicity of formaldehyde and hydrogen chloride in rats. Toxicol. Appl. Pharmacol. 81:401-406.

Siemiatycki, J., ed. 1991. Risk Factors for Cancer in the Workplace. Boca Raton, FL: CRC Press.

Stavert, D.M., D.C. Archuleta, M.J. Behr, and B.E. Lehnert. 1991. Relative acute toxicities of hydrogen fluoride, hydrogen chloride, and hydrogen bromide in nose- and pseudo-mouth-breathing rats. Fundam. Appl. Toxicol. 16:636-655.

Steenland, K., T. Schnorr, J. Beaumont, W. Halperin, and T. Bloom. 1988. Incidence of laryngeal cancer and exposure to acid mists. Br. J. Ind. Med. 45:766-776.

Stevens, B., J.Q. Koenig, V. Rebolledo, Q.S. Hanley, and D.S. Covert. 1992. Respiratory effects from the inhalation of hydrogen chloride in young adult asthmatics. JOM 34:923-929.

ten Berge, W.F., A. Zwart, and L.M. Appleman. 1986. Concentration-time mortality response relationship of irritant and systemically acting vapours and gases. J. Hazard. Mater. 13:301-309.

Toxigenics, Inc. 1984. 90-day Inhalation Toxicity Study of Hydrogen Chloride Gas in B6C3F1 Mice, Sprague-Dawley Rats, and Fischer-344 Rats, Revised. Decatur, IL: Toxigenics, Inc. Pp. 66.

Turlo, S.M., and I. Broder. 1989. Irritant-induced occupational asthma. Chest 96:297-300.

Vernot, E.H., J.D. MacEwen, C.C. Haun, and E.R. Kinkead. 1977. Acute toxicity and skin corrosion data for some organic and inorganic compounds and aqueous solutions. Toxicol. Appl. Pharmacol. 42:417-423.

Wohlslagel, J., L.C. DiPasquale, and E.H. Vernot. 1976. Toxicity of solid rocket motor exhaust: Effects of HCl, HF, and alumina on rodents. J. Combust. Toxicol. 3:61-70.

APPENDIX A

Time-Scaling Calculations for Hydrogen Chloride

Derivation of AEGL-1

Key study: Stevens et al. 1992

Toxicity end point: No-observed-adverse-effect level in exercising asthmatic subjects.

Time-scaling: $C^1 \times t = k$ (ten Berge 1986);
$(1.8\ \text{ppm})^1 \times 0.75\ \text{h} = 1.35\ \text{ppm·h}$

Uncertainty factor: None

10- min, 30-min , 1-h, 4-h, and 8-h AEGL-1: 1.8 ppm

Derivation of AEGL-2

10-min AEGL-2

Key study: Barrow et al. 1977

Toxicity end point: Mouse RD_{50} of 309 ppm to obtain a concentration causing irritation.

10-min AEGL-2: 309 ppm ÷ 3 = 100 ppm

30-min, 1-, 4-, and 8-h AEGL-2

Key Study: Stavert et al. 1991

Toxicity end point: Severe nasal (nose-breathers) or pulmonary (mouth-breathers) effects in rats exposed at 1,300 ppm for 30 min.

Time-scaling: $C^1 \times t = k$ (ten Berge 1986);
$(1{,}300 \text{ ppm})^1 \times 0.5 \text{ h} = 650 \text{ ppm·h}$

Uncertainty factors: 3 for intraspecies variability
3 for interspecies variability

Modifying factor: 3 for sparse database

30-min AEGL-2: $C^1 \times 0.5 \text{ h} = 650 \text{ ppm·h}$
$C = 1{,}300$ ppm
30 min AEGL-2 = 1,300 ppm ÷ 30 = 43 ppm

1-h AEGL-2: $C^1 \times 1 \text{ h} = 650 \text{ ppm·h}$
$C = 650$ ppm
1-h AEGL-2 = 650 ppm ÷ 30 = 21.6 ppm

4-h AEGL-2: 1-h AEGL-2 ÷ 2 = 11 ppm

8-h AEGL-2: 1-h AEGL-2 ÷ 2 = 11 ppm

Derivation of AEGL-3

Key Study: Wholslagel et al. 1976; Vernot et al. 1977

Toxicity end point: One-third of the rat 1-h LC_{50} as an estimate of a no-effect level for death (3,124 ppm ÷ 3 = 1,041 ppm)

Time-scaling: $C^1 \times t = k$ (ten Berge 1986)
$(1{,}041 \text{ ppm})^1 \times 1 \text{ h} = 1{,}041 \text{ ppm·h}$

Uncertainty factors: 3 for intraspecies variability
3 for interspecies variability

10-min AEGL-3: $C^1 \times 0.167\ \text{h} = 1{,}041\ \text{ppm·h}$
$C = 6{,}234\ \text{ppm}$
10-min AEGL-3 = 6,234 ppm ÷ 10 = 623.4 ppm

30-min AEGL-3: $C^1 \times 0.5\ \text{h} = 1{,}041\ \text{ppm·h}$
$C = 2{,}082\ \text{ppm}$
30-min AEGL-3 = 2,082 ppm ÷ 10 = 208 ppm

1-h AEGL-3: $C^1 \times 1\ \text{h} = 1{,}041\ \text{ppm·h}$
$C = 1{,}041\ \text{ppm}$
1-h AEGL-3 = 1,041 ppm ÷ 10 = 104.1 ppm

4-h AEGL-3: $C^1 \times 4\ \text{h} = 1{,}041\ \text{ppm·h}$
$C = 260.25\ \text{ppm}$
4-h AEGL-3 = 260.25 ppm ÷ 10 = 26 ppm

8-h AEGL-3: 8-h AEGL-3 = 4-h AEGL-3 = 26 ppm

APPENDIX B

ACUTE EXPOSURE GUIDELINE LEVELS FOR HYDROGEN CHLORIDE (CAS Reg. No. 7647-01-0)

DERIVATION SUMMARY

AEGL-1				
10 min	30 min	1 h	4 h	8 h
1.8 ppm	1.8 ppm	1.8 ppm	1.8 ppm	1.8 ppm
Key reference: Stevens, B. et al. 1992. Respiratory effects from the inhalation if hydrogen chloride in young adult asthmatics. JOM. 34: 923-929.				
Test species/strain/number: human/adult asthmatic subjects/10				
Exposure route/concentrations/durations: inhalation at 0, 0.8, or 1.8 ppm for 45 min while exercising (1.8 ppm was determinant for AEGL-1)				
Effects: No treatment-related effects were observed in any of the individuals tested				
End point/concentration/rationale: The highest concentration tested was a no-effect level for irritation in a sensitive human population (10 asthmatic individuals tested) and was selected as the basis for AEGL-1. Effects assessed included sore throat, nasal discharge, cough, chest pain or burning, dyspnea, wheezing, fatigue, headache, unusual taste or smell, total respiratory resistance, thoracic gas volume at functional residual capacity, forced expiratory volume, and forced vital capacity. All subjects continued the requisite exercise routine for the duration of the test period.				
Uncertainty factors/rationale: Interspecies: 1, test subjects were human Intraspecies: 1, test subjects were sensitive population (exercising asthmatic subjects)				
Modifying factor: Not applicable				
Animal to human dosimetric adjustment: Insufficient data				
Time-scaling: The AEGL-1 values for a sensory irritant were held constant across time because it is a threshold effect and prolonged exposure will not result in an enhanced effect. In fact one may become desensitized to the respiratory tract irritant over time. Also, this approach was considered valid since the end point (no treatment-related effects at the highest concentration tested in exercising asthmatics) is inherently conservative				

AEGL-1 *Continued*
Data quality and research needs: The key study was well conducted in a sensitive human population and is based on no treatment-related effects. In addition, the direct-acting irritation response is not expected to vary greatly among individuals. Therefore, confidence in the AEGL values derived is high.

AEGL-2				
10 min	30 min	1 h	4 h	8 h
100 ppm	43 ppm	22 ppm	11 ppm	11 ppm
Key references: Stavert et al. 1991. Relative acute toxicities of hydrogen chloride, hydrogen fluoride, and hydrogen bromide in nose- and pseudo-mouth-breathing rats. Fundam. Appl. Toxicol. 16: 636-655. (30-min, 1-, 4-, and 8-h) Barrow, C.S., Alarie, Y., Warrick, M., and Stock, M.F. 1977. Comparison of the sensory irritation response in mice to chlorine and hydrogen chloride. Arch. Environ. Health. 32:68-76. (10-min)				
Test species/strain/number: F-344 rats, 8 males/concentration (30-min, 1-, 4-, and 8-h); Male Swiss Webster mice (10-min)				
Exposure route/concentrations/durations: inhalation at 0 or 1,300 ppm for 30 min (1,300 ppm was determinant for 30-min, 1-, 4-, and 8-h AEGL-2)				
Effects (30-min, 1-, 4-, and 8-h): 0 ppm, no effects; 1,300 ppm, severe necrotizing rhinitis, turbinate necrosis, thrombosis of nasal submucosa vessels in nose-breathers; 1,300 ppm, severe ulcerative tracheitis accompanied by necrosis and luminal ulceration in mouth-breathers (determinant for AEGL-2); RD_{50} = 309 ppm (determinant for 10-min AEGL-2)				
End point/concentration/rationale: 1,300 ppm for 30 min, severe lung effects (ulcerative tracheitis accompanied by necrosis and luminal ulceration) or nasal effects (necrotizing rhinitis, turbinate necrosis, thrombosis of nasal submucosa vessels histopathology) in pseudo-mouth-breathing male F-344 rats (30-min, 1-, 4-, and 8-hr); RD_{50} of 309 ppm ÷ 3 to estimate irritation (10-min)				
Uncertainty Factors/Rationale (30-min, 1-, 4-, and 8-hr): Total uncertainty factor: 10 Intraspecies: 3, steep concentration-response curve implies limited individual variability Interspecies: 3, the use of an intraspecies uncertainty factor of 10 would bring the total uncertainty/modifying factor to 100 instead of 30. That would generate AEGL-2 values that are not supported by data on exercising asthmatic subjects, an especially sensitive subpopulation because exercise increases hydrogen chloride uptake and exacerbates irritation. No effects were noted in exercising young adult asthmatic subjects exposed to HCl at 1.8 ppm for 45 min (Stevens et al. 1992). Using a total UF of 30 would yield 4- and 8-h values of 3.6 ppm (instead of 11 ppm). It is not supportable to predict that humans would be disabled by exposure at 3.6 ppm for 4- or 8- h when exercising asthmatic subjects exposed to one-half that level for 45 min had no effects. The				

AEGL-2 *Continued*
shorter time points would yield values 4-7 times above 1.8 ppm; however, the confidence in the time scaling for hydrogen chloride is good for times up to 100-min because the *n* value was derived from a regression analysis of rat and mouse mortality data with exposure durations ranging from 1 min to 100 minutes. The 30-min value of 43 ppm derived with the total UF of 10 is reasonable in light of the fact that baboons exposed to 500 ppm for 15 min experienced only a slightly increased respiratory rate
Modifying factor: 30-min, 1-, 4-, and 8-h: 3, based on sparse database for AEGL-2 effects and the fact that the effects observed at the concentration used as the basis for AEGL-2 were somewhat severe 10-min: the 10-min AEGL-2 was derived by dividing the mouse RD_{50} of 309 ppm by a factor of 3 to obtain a concentration causing irritation (Barrow et al. 1977). One-third of the mouse RD_{50} for hydrogen chloride corresponds to an approximate decrease in respiratory rate of 30%, and decreases in the range of 20-50% correspond to moderate irritation (ASTM 1991).
Animal to human dosimetric adjustment: Insufficient data
Time-scaling: $C^n \times t = k$ where $n = 1$, based on regression analysis of combined rat and mouse LC_{50} data (1 min to 100 min) reported by ten Berge et al. (1986). Data point used to derive AEGL-2 was 30 min. AEGL-2 values for 1-h exposure period was based on extrapolation from the 30-min value. The 4- and 8-h AEGL-2 values were derived by applying a modifying factor of 2 to the 1-h AEGL-2 value because time scaling would yield a 4-h AEGL-2 value of 5.4 ppm and an 8-h AEGL-2 of 2.7 ppm, close to the 1.8 ppm tolerated by exercising asthmatic subjects without adverse health effects.
Data quality and research needs: Confidence is moderate since the species used is more sensitive than primates to the effects of hydrogen chloride, the chemical is a direct-acting irritant, and a modifying factor was included to account for the relative severity of effects and sparse database.

AEGL-3				
10 min	30 min	1 h	4 h	8 h
620 ppm	210 ppm	100 ppm	26 ppm	26 ppm

Key references: Vernot, E.H., MacEwen, J.D., Haun, C.C., Kinkead, E.R. 1977. Acute toxicity and skin corrosion data for some organic and inorganic compounds and aqueous solutions. Toxicol. Appl. Pharmacol. 42: 417-423.
Wohlslagel, J., DiPasquale, L..C., Vernot, E.H. 1976. Toxicity of solid rocket motor exhaust: Effects of HCl, HF, and alumina on rodents. J. Combustion Toxicol. 3: 61-70.

Test species/strain/gender/number: Sprague-Dawley rats, 10 males/concentration

Exposure route/concentrations/durations: inhalation at 0, 1,813, 2,585, 3,274, 3,941, or 4,455 ppm for 1 h

Effects:

Concentration	*Mortality*
0 ppm	0/10
1,813 ppm	0/10
2,585 ppm	2/10
3,274 ppm	6/10
3,941 ppm	8/10
4,455 ppm	10/10

LC_{50} reported as 3,124 ppm (determinant for AEGL-3)

End point/concentration/rationale: one-third of the 1-h LC_{50} (1,041 ppm) was the estimated concentration causing no deaths.

Uncertainty Factors/Rationale:
Total uncertainty factor: 10

Intraspecies: 3, steep concentration-response curve implies limited individual variability

Interspecies: 3, because (1) the steep concentration-response curve for lethality observed in the Wohlslagel et al. (1976) study in which 1,041 ppm (one-third of the LC_{50} of 3124 ppm) was lower than the LC_0 of 1,813 ppm. This is a conservative selection of the LC_0 and the steep concentration-response curve argues for little interindividual variability; (2) AEGL-3 values generated from a total uncertainty factor of 30 would be close to the AEGL-2 values (within a factor of 2) generated above which are reasonable when compared with data on exercising asthmatics; (3) Sellakumar et al. (1985) exposed rats to HCl at 10 ppm for 6 h/d, 5 d/wk for life and only observed increased trachael and laryngeal hyperplasia. The estimated 6-h AEGL-3 using an intraspecies uncertainty factor of 3 is 17 ppm, close to the level used in the lifetime

AEGL-3 *Continued*
study in which only mild effects were induced; (4) rats exposed at 50 ppm for 6 h/d, 5 d/wk for 90 d (Toxigenics 1984) exhibited mild rhinitis. This level is already 2 times that of the AEGL-3 value for death. Thus, the total uncertainty factor is 10.
Modifying factor: Not applicable
Animal to human dosimetric adjustment: Insufficient data
Time scaling: $C^n \times t = k$ where n = 1, based on regression analysis of rat and mouse mortality data (1 min to 100 min) reported by ten Berge et al. (1986). Reported 1-h data point was used to derive AEGL-3 values. AEGL-3 values for 10-min, 30-min, and 4-h were based on extrapolation from the 1-h value. The 4-h value was adopted as the 8-h value.
Data quality and research needs: Study is considered appropriate for AEGL-3 derivation because exposures are over a wide range of HCl concentrations and utilize a sufficient number of animals. Data were insufficient to derive a no-effect level for death. One-third of the LC_{50} has been utilized previously for chemicals with steep concentration-response curves. Also, in the key study, no deaths were observed in rats exposed at 1,813 ppm.

3

Hydrogen Fluoride[1]

Acute Exposure Guideline Levels

SUMMARY

Hydrogen fluoride (HF) is a colorless, highly irritating, corrosive gas. Reaction with water is rapid, producing heat and hydrofluoric acid. HF is used in the manufacture of artificial cryolite; in the production of aluminum, fluorocarbons, and uranium hexafluoride; as a catalyst in alkylation processes during petroleum refining; in the manufacture of fluoride salts; and in stainless-steel pickling operations. It is also used to etch glass and as a cleaner in metal finishing processes.

HF is a severe irritant to the eyes, skin, and nasal passages; high concentrations may penetrate to the lungs, resulting in edema and hemorrhage. Data on irritant effects in humans and lethal and sublethal effects in six species of mammal (monkey, dog, rat, mouse, guinea pig, and rabbit) were available for developing acute exposure guideline levels (AEGLs). The

---

[1]This document was prepared by the AEGL Development Team comprising Sylvia Talmage (Oak Ridge National Laboratory) and National Advisory Committee (NAC) on Acute Exposure Guideline Levels for Hazardous Substances member Larry Gephart (Chemical Reviewer). The NAC reviewed and revised the document and the AEGL values as deemed necessary. Both the document and the AEGL values were then reviewed by the National Research Council (NRC) Subcommittee on Acute Exposure Guideline Levels. The NRC subcommittee concluded that the AEGLs developed in this document are scientifically valid conclusions on the basis of the data reviewed by the NRC and are consistent with the NRC guidelines reports (NRC 1993, 2001).

data were considered adequate for deriving the three AEGL classifications for the five exposure periods. Regression analyses of the reported concentration-exposure durations for lethality in the animal species determined that the relationship between concentration and time is $C^2 \times t = k$ (where C = concentration, t = time, and k is a constant).

The AEGL-1 was based on an exposure at 3 parts per million (ppm) (range, 0.85-2.93 ppm) for 1 hour (h), which was the threshold for pulmonary inflammation, as evidenced by an increase in the percentage of several inflammatory parameters such as CD3 cells and myeloperoxidase in the bronchoalveolar lavage fluid of 20 healthy exercising adult subjects (Lund et al. 1999). There were no increases in neutrophils, eosinophils, protein, or methyl histamine at this or the next higher average exposure concentration of 4.7 ppm (range, 3.05-6.34 ppm). There were no changes in lung function and only minor symptoms of irritation at that concentration (Lund et al. 1997). Although healthy adults were tested, several individuals had increased immune factors, indicating atopy. The 3-ppm concentration was divided by an intraspecies uncertainty factor (UF) of 3 to protect susceptible individuals. Because there were no effects on respiratory parameters of healthy adults at concentrations up to 6.34 ppm in the Lund et al. (1997) study and at concentrations up to 8.1 ppm for 6 h/day (d) with repeated exposures in a supporting study (Largent 1960, 1961), the calculated AEGL-1 values will be protective of asthmatic individuals. Although the Lund et al. (1999) study duration was only 1 h, the longer exposures at higher concentrations in the supporting study (Largent 1960, 1961), and the fact that adaptation to mild sensory irritation occurs, support application of the 1-ppm concentration for up to 8 h.

The 10-minute (min) AEGL-2 was based on an absence of serious pulmonary or other adverse effects in rats during direct delivery of HF to the trachea at 950 ppm for an exposure period of 10 min (Dalbey 1996; Dalbey et al. 1998a). The reported concentration-exposure value of 950 ppm for 10 min was adjusted by a combined UF of 10—3 for interspecies variation, because the rat was not the most sensitive species in other studies (but direct delivery to the trachea is a sensitive model), and an intraspecies UF of 3 to protect susceptible individuals. The resulting 10-min value clearly is below the serious injury categories of data from tests in monkeys, rats, dogs, mice, guinea pigs, and rabbits.

The 30-min and 1-, 4- and 8-h AEGL-2 values were based on a study in which dogs exposed at 243 ppm for 1 h exhibited blinking, sneezing, and coughing (Rosenholtz et al. 1963). Rats exposed at a similar concentration (291 ppm) developed moderate eye and nasal irritation. The next higher concentration (489 ppm for 1 h) resulted in respiratory distress and severe eye and nasal irritation in the rat, signs more severe than those ascribed to AEGL-2. The moderate eye and nasal irritation observed in dogs at 243 ppm was considered the threshold for impaired ability to escape. The 1-h value of 243 ppm was adjusted by a total UF of 10—3 for interspecies variation, because the dog is a sensitive species for sensory irritation, and 3 to protect susceptible individuals. The values were scaled across time using $C^n \times t = k$, where $n = 2$. The n value was derived using concentration-exposure duration relationships from animal lethality studies. It should be noted that the resulting 30-min AEGL-2 of 34 ppm is similar to the 32-ppm concentration that could be tolerated by human subjects for only minutes in the Machle et al. (1934) study. Using a larger total UF such as 30 would reduce the 1-h value to 8 ppm, a concentration that resulted in only slight irritation in healthy adults during repeated, intermittent exposures (Largent 1960, 1961). Because the time-scaled 8-h value of 8.6 ppm was inconsistent with the Largent (1960, 1961) study in which humans subjects inhaling 8.1 ppm intermittently suffered no effects other than slight irritation, the 8-h AEGL-2 was set equal to the 4-h AEGL-2.

The 10-min AEGL-3 was based on the reported 10-min lethal threshold of 1,764 ppm reported in orally cannulated rats (Dalbey 1996; Dalbey et al. 1998). That value was rounded to 1,700 ppm and adjusted by UFs of 3 for interspecies differences (LC_{50} values [concentrations lethal to 50% of subjects] differ by a factor of approximately 2-4 between the mouse and rat) and 3 to protect susceptible individuals. The total UF for the 10-min AEGL-3 was 10. Application of a larger UF would reduce the 10-min AEGL-3 to a value below the 10-min AEGL-2.

The 30-min and 1-, 4-, and 8-h AEGL-3 values were derived from a 1-h exposure that resulted in no deaths in mice (Wohlslagel et al. 1976). The data indicated that 263 ppm was the threshold for lethality. A comparison of LC_{50} values among species indicated that the mouse was the most sensitive species in the lethality studies. The 1-h value of 263 ppm was adjusted by an interspecies UF of 1, because the mouse was the most sensitive species, and an intraspecies UF of 3 to protect susceptible individuals. A modifying factor (MF) of 2 was applied to account for the fact that the highest nonlethal value was close to the LC_{50} of 342 ppm. The resulting value was scaled to the other AEGL-specified exposure periods using $C^n \times t = k$, where $n = 2$. A total factor of 6 is reasonable and sufficient, because application of a total factor of 20 (3 each for inter- and intraspecies uncertainties and 2 as a MF) would reduce the predicted 6-h AEGL-3 to 5.4 ppm, a concentration below the peak 8.1-ppm concentration that produced only irritation in humans (Largent 1960, 1961). Because HF is well scrubbed at low concentrations, and because the time-scaled 8-h AEGL-3 value of 15 ppm was inconsistent with data from repeated exposures in animal studies, the 8-h value was set equal to the 4-h value.

The AEGLs for HF are summarized in Table 3-1.

1. INTRODUCTION

HF is a colorless, highly irritating, corrosive gas with a molecular weight of 20.01 and a density of 1.27. It is extremely soluble in water; reaction with water produces heat and forms hydrofluoric acid. At atmospheric pressure, the gas is monomeric; at higher pressures, polymerization takes place, producing a gas of density greater than monomeric HF (Perry et al. 1994). Although HF is lighter than air and would disperse when released, a cloud of vapor and aerosol that is heavier than air may be formed under some release conditions (EPA 1993). Additional chemical and physical properties are listed in Table 3-2.

Anhydrous HF is manufactured and used in the United States for the production of aluminum, fluorocarbons, cryolite, and uranium hexafluoride; in solutions used for glass etching, cleaning, stainless steel pickling, and chemical derivatives; as a catalyst for the production of gasoline; and for nuclear applications (EPA 1993; Perry et al. 1994).

Recent production data were not located. In 1992, HF was manufactured in the United States by three companies at 10 sites with a total production capacity of 206,000 tons; U.S. production is approximately 90% of capacity. In addition, several aluminum producers make HF for on-site use. In 1991, users and/or producers of HF included 13 fluorocarbon production facilities and approximately 51 petroleum refineries that had HF alkylation units. Due to the phase-out of chlorofluorocarbon production, HF production was expected to fall slightly by 1996 (EPA 1993).

Contact of liquid HF with the skin can produce severe burns; the gas is corrosive to the eyes and mucous membranes of the respiratory tract. The acute inhalation toxicity of HF has been studied in several laboratory animal species, and its irritant properties have been studied in human volunteers. Large differences in the concentrations causing the same effects in animal studies indicate that difficulties in measurement techniques were encountered by investigators in some of the early studies, thus limiting the value of their quantitative data. In addition, experimental details and descriptions of effects were inadequate in some of the studies.

TABLE 3-1 Summary Table of AEGL Values (ppm [mg/m^3])

Classification	10 min	30 min	1 h	4 h	8 h	End Point (Reference)
AEGL-1 (Nondisabling)	1.0 (0.8)	1.0 (0.8)	1.0 (0.8)	1.0 (0.8)	1.0 (0.8)	Threshold, pulmonary inflammation in humans (Lund et al. 1997, 1999)
AEGL-2 (Disabling)	95 (78)	34 (28)	24 (20)	12 (9.8)	12 (9.8)	NOAEL for lung effects in cannulated rats (Dalbey 1996; Dalbey et al. 1998a);[a] sensory irritation in dogs (Rosenholtz et al. 1963)[b]
AEGL-3 (Lethal)	170 (139)	62 (51)	44 (36)	22 (18)	22 (18)	Lethality threshold in cannulated rats (Dalbey 1996; Dalbey et al. 1998a);[c] lethality threshold in mice (Wohlslagel et al. 1976)[d]

[a]10-min AEGL-2 value.
[b]30-min and 1-, 4-, and 8-h AEGL-2 values.
[c]10-min AEGL-3 value.
[d]30-min and 1-, 4-, and 8-h AEGL-3 values.
Abbreviations: mg/m^3, milligrams per cubic meter; ppm, parts per million.

2. HUMAN TOXICITY DATA

2.1. Acute Lethality

No data were located regarding human deaths following inhalation-only exposure to HF. However, several studies indicate that humans have died from accidental exposure to hydrofluoric acid (Kleinfeld 1965; Tepperman 1980; Braun et al. 1984; Mayer and Gross 1985; Chan et al. 1987; Chela et al. 1989; ATSDR 1993). These accidents involved acute inhalation of HF in combination with dermal exposure involving severe dermal lesions. Deaths were attributed to pulmonary edema and cardiac arrhythmias, the latter a result of acidosis from pronounced hypocalcemia

TABLE 3-2 Chemical and Physical Data for Hydrogen Fluoride

Parameter	Value	Reference
Synonyms	Hydrofluoric acid gas, anhydrous hydrofluoric acid	Budavari et al. 1996
Molecular formula	HF	Budavari et al. 1996
Molecular weight	20.01	Budavari et al. 1996
CAS Registry Number	7664-39-3	Budavari et al. 1996
Physical state	Gas	Budavari et al. 1996
Color	Colorless	Budavari et al. 1996
Solubility in water	Miscible in all proportions	Perry et al. 1994
Vapor pressure	760 mm Hg at 20°C	ACGIH 2002
Density (water = 1)	1.27 at 34°C	Perry et al. 1994
Melting point	-87.7°C	Perry et al. 1994
Flammability	Not flammable	Weiss 1980
Boiling point	19.5°C	Perry et al. 1994
Conversion factors	1 ppm = 0.82 mg/m^3 1 mg/m^3 = 1.22 ppm	ACGIH 2002

and hypomagnesemia following dermal fluoride uptake. No doses or exposure levels could be determined.

2.2. Nonlethal Toxicity

Ronzani (1909) and Machle et al. (1934) cite early reports in which a concentration of HF at 0.004% (40 ppm) was used in the treatment of tuberculosis. No exposure times were stated. The sharp, irritating odor of HF is noticeable at 0.02-0.13 ppm (Sadilova et al. 1965; Perry et al. 1994).

Three groups of investigators studied the irritant effects of acute HF exposures in human volunteers. An additional study reported on exposures over a period of 10-50 d. Studies of industrial exposures and accidental releases were located, but exposure concentrations either were intermittent or were not measured; furthermore, those studies were confounded by the presence of other chemicals.

2.2.1. Experimental Studies

The studies using human volunteers are summarized in Table 3-3. Machle et al. (1934) exposed two male volunteers to concentrations of HF at 0.1, 0.05, and 0.026 mg/L (32, 61, and 122 ppm) for very short exposure periods. Inhalation of HF at 122 ppm produced marked conjunctival and respiratory irritation within 1 min and smarting of the exposed skin. At 61 ppm, eye and nasal irritation were marked, but smarting of the skin was not reported. Irritation of the eyes and nose was mild at 32 ppm, and that concentration was "tolerated" with discomfort. At all concentrations, irritation of the larger airways and a sour taste in the mouth were present. Repeated exposures (undefined) failed to produce adaptation.

Collings et al. (1951) subjected two volunteers to an atmosphere containing HF and silicon tetrafluoride during an 8-h work shift; the subjects left the area for 15 min every 2 h and during a lunch break. The average concentration of fluoride during the exposure was 3.8 milligrams per cubic meter (mg/m^3) (4.6 ppm); the concentration of HF alone was not measured, but would presumably have been $\leq$4.6 ppm. According to the authors, "both subjects experienced the anticipated irritant effect of thegases and the remarkably rapid acclimation which is so well known." No further details on irritant effects were stated.

Largent (1960, 1961) exposed five male volunteers (ages 17-46) to variable concentrations of HF for 6 h/d over a period of 10-50 d. Average individual concentrations over the exposure period ranged from 1.42 ppm to 4.74 ppm (average, 3.2 ppm; total range, 0.9-8.1 ppm). Effects were no more severe in two subjects who were exposed at concentrations up to 7.9 ppm and up to 8.1 ppm over a 25 d and 50 d period, respectively, than in the other subjects. Although it was stated that one subject tolerated 1.42 ppm for 15 d (6 h/d) without noticeable effects, exposure of the same subject at 3.39 ppm for 10 d at a later time resulted in redness of the face and, by day 11, some flaking of the skin. The subjects experienced very slight irritation of the eyes, nose, and skin at $\leq$2 ppm and noted a sour taste in the mouth during the exposures. It is not clear whether the subject exposed at 1.42 ppm for 15 d also experienced those effects. Application of a coating of face cream prior to exposure was found to prevent any discomfort or redness of the shaved facial skin. Any signs of discomfort disappeared after cessation of exposure. Systemic effects were not observed. Two subjects in this study displayed slightly different levels of sensitivity. One subject suffered from a cold for a few days during which there was heightened discomfort. Another subject did not use cosmetic cream.

TABLE 3-3 Summary of Sensory and Irritant Effects in Humans

Concentration (ppm)	Exposure Time	Effects	Reference
0.02-0.13	NA	Odor threshold	Perry et al. 1994; Amoore and Hautala 1983; Sadilova et al. 1965
0.2-0.7	1 h	No to low sensory and upper airway irritation; no change in FEV_1, decrease in FVC; no change in components of BAL	Lund et al. 1997, 1999
0.85-2.9	1 h	No to low sensory and upper airway irritation; no change in FVC, FEV_1; BAL showed increase in CD3 cells, lymphocytes, with no increase in neutrophils, eosinophils, protein	Lund et al. 1997, 1999
3.0-6.3	1 h	No eye irritation, but upper (3/14 subjects) and lower (1/14 subjects) respiratory airway irritation;[a] no change in FVC, FEV_1; BAL showed increase in CD3 cells, lymphocytes, myeloperoxidase, cytokine, with no increase in neutrophils, eosinophils, protein	Lund et al. 1997, 1999
1.42	6 h/d, 15 d	No noticeable effect (single subject)	Largent 1960, 1961
2.59-4.74 (average) 0.9-8.1 (range)	6 h/d, 10-50 d	Slight irritation of the skin, nose, and eyes; sour taste in mouth	Largent 1960, 1961
4.6 (average)[b] 3.5-7.1 (range)	7 h	Irritant effect followed by adaptation	Collings et al. 1951

32	3 min	"Tolerated" with discomfort; mild irritation of eyes and nose	Machle et al. 1934
61	Approx. 1 min	Eye and nasal irritation	Machle et al. 1934
122	Approx. 1 min	Marked eye and respiratory irritation, skin irritation, highest concentration tolerated for >1 minute	Machle et al. 1934

[a]Upper airways: symptoms of eye, nose and throat irritation; lower airways: symptoms of chest tightness, coughing, expectoration, wheezing.
[b]Exposure to gaseous HF and silicon tetrafluoride; value expressed as fluoride ion.
Abbreviations: BAL, bronchoalveolar lavage fluid; FEV_1, forced expiratory volume in one second; FVC, forced vital capacity; NA, not applicable.

Lund et al. (1995) exposed 15 healthy male volunteers to concentrations at 1.5-6.4 mg/m^3 (1.83-7.8 ppm) for 1 h in order to study sensory irritation, as indicated by inflammatory cells in bronchoalveolar lavage (BAL) fluid and changes in pulmonary parameters. HF induced a bronchial inflammatory reaction as indicated by an increase in the fraction of lymphocytes and neutrophils in the BAL fluid. The fraction of CD5 positive cells (cluster determinants, a subpopulation of T cells) increased from a pre-exposure value of 0.6% to 6.3% at 20-24 h after exposure. There were no changes in spirometry measurements. The data were reported in an abstract, and no further details were given.

In a more recent publication that appears to be a continuation of the above study, 20 healthy, nonsmoking male volunteers, ages 21-44 y, were exposed to a constant concentration within the range of 0.24-6.34 ppm for 1 h in a 19.2-m^3 chamber (Lund et al. 1997). Three subjects per exposure group were exposed twice with a 3-month (mo) interval between exposures. In order to analyze a dose-response relationship, the exposures were divided into ranges of 0.2-0.7 ppm (nine subjects); 0.85-2.9 ppm (seven subjects); and 3.0-6.3 ppm (seven subjects). Two of the subjects had hay fever; one of those and an additional subject had an increased total IgE immunoglobulin level. The exposure groups of these subjects were not identified. The authors stated that the rest of the subjects were not atopic or allergic. Exposure concentrations were monitored by an electrochemical sensor. Exact exposures were measured by collecting air samples on cellulose pads impregnated with sodium formate and analyzed with a fluoride selective electrode. Upper and lower airway and eye irritation were subjectively scored on a scale of 0 (no symptoms) to 5 (severe symptoms). In addition, FEV_1 (forced expiratory volume in one second) and FVC (forced vital capacity) were measured before, during (every 15 min), and at the end of the exposures and again at 4 and 24 h post-exposure. Subjects rested during the first 45 min of exposure; during the last 15 min the subjects exercised on a stationary bicycle.

Five subjects reported minor upper and lower respiratory symptoms (mild coughing or expectoration and itching of the nose) before entering the chamber. Symptoms increased after the 1-h exposure, but none of the subjects in the lower two exposure groups reported symptom scores of greater than 3. Specific scores for symptoms were not reported in the publication; however, a score of 1-3 was defined as low. The mean FVC was significantly decreased after exposure in the lowest exposure group, from 5.1 liters (L) to 4.8 L. The lack of significant changes in the higher exposure groups makes it unlikely that the change in FVC in the lowest group was a result of chemical exposure. In the highest exposure group, no eye irritation was reported, but three subjects reported upper airway irritation (itching or soreness of the nose or throat) with scores of greater than 3, and one subject reported a lower airway irritation (chest tightness, soreness, coughing, expectoration, or wheezing) with a score of greater than 3. Specific symptoms and actual scores were not reported. The authors noted that lower airway symptoms were not reported to a significant degree in relation to exposure to HF, and none of the subjects had obvious signs of bronchial constriction. The authors note that the study was not blind and that the symptoms may have been overreported; however, the exposed subjects were unaware of the exposure concentration.

In a second publication addressing the same study (Lund et al. 1999), the authors reported whether or not changes in BAL fluid components occurred 24 h after 1-h exposures at the above concentrations. In particular, they looked at an inflammatory response as indicated by changes in types of white blood cells and several noncellular components compared with measurements taken 3 weeks (wk) before the exposures. The aspirated BAL was divided into bronchial and bronchoalveolar portions, the latter reflecting the more distal air spaces of the lung. Results were provided in the form of cell differentials (%, median and interquartile ranges), making absolute comparisons difficult. The percentage of CD3-positive cells was significantly increased in the bronchial portions of the BAL in the two higher exposure groups and in the bronchoalveolar portions of the BAL in the highest exposure group (3.0-6.3 ppm). CD (cluster determinant) cells are a subpopulation of T cells (i.e., lymphocytes from the thymus) that are recognizable by a selective monoclonal antibody. Although neutrophils were not increased, myeloperoxidase and interleukin-6 (a cytokine) increased significantly in the bronchial portion in the highest exposure group. There were no dose-response related differences in percentages of lymphocytes, eosinophils, neutrophils, and macrophages among the groups for either portions of the BAL, although for the exposure groups combined, the percentage of lymphocytes increased slightly but significantly and the percentage of macrophages decreased slightly but significantly compared with pre-exposure values in both portions of the BAL. Methyl histamine and intercellular adhesion molecule-1 in the bronchial portion were unchanged and, surprisingly, several protein components, including albumin and total protein, were decreased in the bronchoalveolar portion. Although the authors refer to an inflammatory response, they considered the effects minor and could not identify a clear concentration-response relationship. Increases in neutrophils, eosinophils, mast cells, or serum protein in the

BAL are considered biomarkers of inflammation (NRC 1989); there were no concentration-related increases in any of those BAL components in the study.

2.2.2. Worker Exposure

Chronic exposures in industrial situations have led to skeletal fluorosis in exposed workers. Concentrations of airborne HF in those studies are often estimated or unknown, and exposures are usually to both HF and fluoride dusts (NIOSH 1976; ATSDR 1993). However, studies with long-term exposure levels can be used to determine no-effect concentrations. For example, Derryberry et al. (1963) reported that there were no statistically significant differences in several respiratory parameters between a control group and a group of 57 workers engaged in the manufacture of phosphate fertilizer. Exposure to dust and HF gas combined resulted in fluoride concentrations ranging from 0.50 mg/m^3 to 8.32 mg/m^3, with an average for the group of 2.81 mg/m^3 (HF at 3.6 ppm) over a 14-y period.

Machle and Evans (1940) studied a group of workers exposed to HF and, to a lesser extent, calcium fluoride dust during the manufacture of hydrofluoric acid. Over a 5-y period, the workers were exposed intermittently, in the vicinity of equipment or while repairs were made, to concentrations of fluoride at 0.011-0.021 mg/L (HF at 14-27 ppm). Medical examinations revealed no clinical or roentgenologic evidence of damage.

A case of chronic poisoning of a worker exposed to HF at an alkylation unit of an oil company was documented by Waldbott and Lee (1978). During his 10 y of almost daily exposure, acute episodes occurred 10-15 times a year. Acute symptoms consisted of intense eye irritation, tearing, blurred vision, marked dyspnea, nausea, epigastric pain, vomiting, and sudden weakness. The worker had repeated minor HF "burns" on the skin. Unfortunately, no monitoring data were available. Estimates of exposure concentrations given by the worker and his coworkers (e.g., a concentration of >25 ppm during acid-tank gauging) are of limited value. During the 10-y period the previously healthy worker suffered increasingly worsening back and leg pains, loss of memory, osteoarthritis, restrictive and obstructive lung disease, and hematuria.

Abramson et al. (1989) cited worker exposures to HF at 0.2-4.1 mg/m^3 (0.2-5 ppm) in aluminum smelter plants. Although asthma and chronic obstructive lung disease appear to be associated with work in aluminum smelters, the confounding factors of multiple chemical exposure, small sample size, and cigarette smoking did not support a causal relationship.

2.2.3. Accidents

Three documented cases of accidental release of HF were located. A fourth accident was cited in an EPA (1993) report. Over a 48-h period, approximately 53,000 lb of anhydrous HF and 6,600 lb of isobutane were released from a petrochemical plant in Texas in October, 1987 (Wing et al. 1991). The nearest residential community was 0.25 miles from the plant. Within 20 min of the release, persons within 0.5 miles of the plant were evacuated; eventually a 5-square-mile area was evacuated (3,000 people). Samples taken downwind (distance not stated) 1 h after the release contained 10 ppm; samples obtained after 2 h contained "minimal traces" of HF. The most frequently reported symptoms in people who presented at emergency rooms at two area hospitals were eye irritation (41.5%), throat burning (21.0%), headache (20.6%), shortness of breath (19.4%), throat soreness (17.5%), chest pain (16.9%), cough (16.4%), and nausea (15%).

Dayal et al. (1992) conducted follow-up evaluations of subjects involved in the Texas HF-exposure incident. Two years after the accident, 10,811 individuals were surveyed. Symptom surveys were completed by 1,994 of the 10,811. Individuals were balanced for gender, age, and predisposition across exposure categories of high, intermediate, none, and discordant (unknown or not well defined). A mathematical model was used to predict isodensity curves of HF concentrations at the time of the accident. However, no concentrations were mentioned in the study. Three symptoms were used for exposure assessment: burning or irritation of the throat, burning or irritation of the eyes, and coughing or difficulty breathing. Symptoms reported immediately after the accident were compared with symptoms reported 2 y later. There was a strong dose relationship between the exposure symptoms reported following the accident and those reported 2 y later. Although substantial improvements in health were apparent 2 y after the accident, some symptoms persisted, notably breathing problems and eye symptoms. The authors discussed the problems of recall bias and behavioral sensitization, which would result in an overestimation of the effects.

In another incident, a cloud of gases was released from an oil refinery near Tulsa, Oklahoma, on March 19, 1988 (Himes 1989). The major constituent of the cloud was HF, which may have reached an airborne concentration of 20 ppm. A total of 36 people, including emergency personnel responding to the incident, were treated at area hospitals for acute chemical exposure. There were no fatalities. No measurements were taken and no further details of the incident were given.

In a third incident, 13 workers at an oil refinery were exposed to hydrofluoric acid mist at a maximum concentration of 150-200 ppm for approximately 2 min (Lee et al. 1993). Prompt treatment with nebulized calcium gluconate was administered. The workers were medically evaluated within an hour of exposure, at which time the only symptoms were minor upper respiratory tract irritation.

EPA (1993) cited a study by Trevino (1991) that described an industrial accident in Mexico that resulted in exposure of seven workers at approximately 10,000 ppm for several minutes. Periodic examinations for up to 11 y after exposure revealed no long-term or delayed effects. No measurement methods and no further details of the study were provided.

2.3. Developmental and Reproductive Toxicity

No studies were located regarding reproductive or developmental effects in humans resulting from inhalation exposure to HF. Fluoride is rapidly absorbed following oral ingestion, crosses the placenta in limited amounts, and is found in placental and fetal tissue (ATSDR 1993). Studies on the incidence of reproductive or developmental effects in areas using fluoridated water have found no correlation between fluoridation levels and birth defects (ATSDR 1993).

2.4. Genotoxicity

No data concerning the genotoxicity of HF in humans were identified in the available literature.

2.5. Carcinogenicity

Although several studies indicated an increase in respiratory cancers among workers in several industries who could be exposed to HF or fluoride dusts, the confounding factors of exposure to other chemicals and smoking status, along with the lack of clear exposure concentrations, make the studies of questionable relevance (ATSDR 1993). The potential carcinogenicity of fluoride is debatable. EPA has not yet evaluated fluoride for potential human carcinogenicity.

2.6. Summary

Four studies with human volunteers reported both measured concentrations and exposure durations. Human volunteers could "tolerate" a concentration at 32 ppm for 3 min, reporting only mild irritation of the eyes and nose (Machle et al. 1934). The highest concentration that could be voluntarily tolerated for more than 1 min was 122 ppm. Irritation was slight in humans during repeated exposures (6 h/d for up to 50 d) at mean concentrations of 2.59-4.74 ppm (range 0.9-8.1 ppm), but not at 1.42 ppm for 15 d. In general, concentrations at ≤2 ppm for 6 h/d were considered only slightly irritating (Largent 1960, 1961). Male subjects (3-4 of 14) reported upper and lower respiratory irritation of >3 on a scale of 0 to 5 at 3.0-6.3 ppm (Lund et al. 1997). None of the subjects had obvious symptoms of bronchial constriction (Lund et al. 1999). No human lethality studies following inhalation-only exposures were located. No data on developmental and reproductive effects, genotoxicity, or carcinogenicity following inhalation exposures were located.

3. ANIMAL TOXICITY DATA

3.1. Acute Lethality

Data on single exposures to HF resulting in mortality are available for the monkey, rat, mouse, guinea pig, and rabbit. Those data are summarized in Table 3-4. Results of a study with an animal model that simulates human mouth-breathing are reported in Table 3-5, and studies using repeated exposures are reported in the text.

3.1.1. Nonhuman Primates

Groups of four male and female rhesus monkeys were exposed to concentrations of HF at 690, 1,035, 1,575, 1,600, 1,750, or 2,000 ppm for 1 h (MacEwen and Vernot 1970). No deaths occurred at 690, 1,575, or 1,600 ppm; one death occurred in the group exposed at 1,035 ppm; and three deaths occurred in both the group exposed at 1,750 ppm and the group exposed at 2,000 ppm. Using probit analysis, the authors calculated a LC_{50} of 1,774 ppm (95% confidence limit, 1,495-2,105). Massive lung hemorrhage and edema were present in animals that died. Signs of toxicity during exposures included respiratory distress, paresis, salivation, lacrimation, nasal

TABLE 3-4 Summary of Acute Lethal Inhalation Data in Laboratory Animals

Species	Concentration (ppm)	Exposure Time	Effect[a]	Reference
Monkey	1,774	1 h	LC_{50}	MacEwen and Vernot 1970
	690	1 h	No deaths	
Rat	25,690	5 min	100% mortality	DiPasquale and Davis 1971;
	18,200	5 min	LC_{50}	MacEwen and Vernot 1971;
	12,440	5 min	10% mortality	Higgins et al. 1972
Rat	4,970	5 min	LC_{50}	Rosenholtz et al. 1963
Rat	2,689	15 min	LC_{50}	Rosenholtz et al. 1963
Rat	2,042	30 min	LC_{50}	Rosenholtz et al. 1963
Rat	2,300	1 h	LC_{50}	Haskell Laboratory 1990
	1,563, 1,827	1 h	LC_{01}	
Rat	2,039	1 h	10% mortality	Dalbey et al. 1998a
	1,224	1 h	No deaths	
Rat	1,395	1 h	LC_{50}	Wohlslagel et al. 1976
	1,108	1 h	20% mortality	
	1,087	1 h	No deaths	
Rat	1,307	1 h	LC_{50}	Rosenholtz et al. 1963
Rat	1,276	1 h	LC_{50}	MacEwen and Vernot 1970
	480	1 h	No deaths	
Rat	966	1 h	LC_{50}	Vernot et al. 1977;
	848	1 h	No deaths	MacEwen and Vernot 1874
Rat	190	6 h	100% mortality	Morris and Smith 1982

Mouse	11,010	5 min	100% mortality	Higgins et al. 1972
	2,430	5 min	No deaths	
Mouse	6,247	5 min	LC_{50}	MacEwen and Vernot 1971;
	2,430	5 min	No deaths	Higgins et al. 1972
Mouse	501	1 h	LC_{50}	MacEwen and Vernot 1970
Mouse	456	1 h	LC_{50}	MacEwen and Vernot 1974;
	351	1 h	No deaths	Vernot et al. 1977
Mouse	342	1 h	LC_{50}	Wohlslagel et al. 1976
	278	1 h	10% mortality	
	263	1 h	No deaths	
Guinea pig	>1,220-1,830	5 min	Death in a significant number of animals	Machle et al. 1934
	1,220	30 min	No deaths, respiratory irritation	
	122	5 h	Injury, no deaths	
Guinea pig	4,327	15 min	LC_{50}	Rosenholtz et al. 1963
	1,377	30 min	No deaths	
Rabbit	>1,220-1,830	5 min	Death in a significant number of animals	Machle et al. 1934
	1,220	30 min	No deaths, respiratory irritation	
	122	5 h	Injury, no deaths	

[a]LC_{50} and 100% mortality values were obtained at 3 h post-exposure (Morris and Smith 1982), 7 d post-exposure (MacEwen and Vernot 1971; Higgins et al. 1972), and 14 d post-exposure (Rosenholtz et al. 1963; MacEwen and Vernot 1970; Wohlslagel et al. 1976; Dalbey et al. 1998a).

TABLE 3-5 Mortality Data in Orally Cannulated Rats[a]

Exposure Duration	Concentration	Mortality (%)
2 min	8,621	5
	4,877	10
	1,589	0
	593	0
10 min	7,014	80
	3,847	50
	1,764	5
	1,454	0
	950	0
	271	0
	135	0
60 min	48	0
	20	0

[a]Animals were exposed via cannula to the trachea.
Sources: Dalbey 1996; Dalbey et al. 1998a.

discharge, gagging, sneezing, and vomiting. Skin burns were observed post-exposure; those healed after several days.

3.1.2. Rats

Groups of 10 young male Wistar rats were exposed to HF at various measured concentrations (concentration range not stated) for 5, 15, 30, or 60 min (Rosenholtz et al. 1963). The survivors were weighed daily and observed for 14 d after exposure. LC_{50} values of 4,970, 2,689, 2,042, and 1,307 ppm, respectively, were calculated. During the exposures, there were signs of irritation of the conjunctiva and nasal passages. They lasted 7 d post-exposure and included reddened conjunctivae, pawing at the nose, marked lacrimation, nasal secretion, and sneezing. In addition to some delayed deaths, respiratory distress, body-weight loss (10-15% during days 3-7 post-exposure), and several days of general weakness also was seen in some animals. After the first 7 d, surviving rats rapidly gained weight and reached a weight level equal to that of controls.

Pathologic examinations were performed on groups of rats exposed in the lethal range for 15 min or 30 min (Rosenholtz et al. 1963). Post-exposure periods ranged from 1 h to 84 d. Gross and microscopic examination revealed concentration-dependent lesions in the kidney, liver, nasal passage, bone marrow, and skin. Those lesions included nasal passage necrosis and associated acute inflammation (the external nares and nasal vestibules turned black), selective renal tubular necrosis, hepatocellular intracytoplasmic globules, dermal collagen changes with acute inflammation, and possible myeloid hyperplasia of the bone marrow. Many of the lesions showed signs of reversibility by 48 h to 7 d after exposure.

Groups of 10 adult Wistar rats were exposed to HF at concentrations ranging from 12,440 ppm to 25,690 ppm for 5 min to calculate the 5-min LC_{50} (DiPasquale and Davis 1971; MacEwen and Vernot 1971; Higgins et al. 1972). Exposure concentrations were continuously monitored using specific ion electrodes. HF produced pulmonary edema of varying degrees of severity in most of the exposed rats. Pulmonary hemorrhage was a common finding in rats that died during or shortly after exposure at concentrations above the LC_{50}. In exposures below the LC_{50}, delayed deaths occurred about 24 h after exposure; occasionally, deaths occurred 3-4 d later. The 5-min LC_{50} was 18,200 ppm. Mortality was 10% at 2,440 ppm and 100% at 25,690 ppm (Higgins et al. 1972).

Groups of eight male Wistar rats were exposed to HF at 480-2,650 ppm for 1 h (MacEwen and Vernot 1970). No deaths occurred at 480 ppm. The LC_{50}, calculated by probit analysis, was 1,276 ppm (95% confidence limit, 1,036-1,566). Massive lung hemorrhage and edema were present in animals that died. Signs of toxicity during exposures included respiratory distress, paresis, salivation, lacrimation, and nasal discharge. In a similar study, groups of five male Sprague-Dawley rats were exposed at 848, 1,097, or 1,576 ppm for 1 h (MacEwen and Vernot 1974; Vernot et al. 1977). No deaths occurred at 848 ppm. The LC_{50} was 966 ppm with 95% confidence limits of 785-1,190 ppm.

Groups of 10 male Sprague-Dawley-derived rats were exposed at 1,087, 1,108, 1,405, 1,565, or 1,765 ppm for 1 h (Wohlslagel et al. 1976). Animals were observed for toxic signs and mortality for up to 14 d post-exposure. Some animals that died following exposure or were sacrificed after the 14-d observation period were examined histologically. The 1-h LC_{50} was 1,395 ppm. Signs during the exposures included eye and mucous membrane irritation, respiratory distress, corneal opacity, and erythema of the exposed skin. Pathologic examinations of rats that died during or after exposure revealed pulmonary congestion, intra-alveolar edema, and some cases of thymic hemorrhage. A concentration of 1,087 ppm was not lethal (0/10 deaths), whereas a concentration of 1,108 ppm resulted in two deaths in 10 subjects.

In another study, exposure of rats to fluoride at 148 mg/m^3 (HF at 190 ppm) for 6 h resulted in 100% mortality within 3 h post-exposure (Morris and Smith 1982). Discharge of fluid from the external nares was observed prior to death, but no lung lesions were present.

Groups of 4 male rats were exposed head-only to various concentrations of anhydrous HF for time periods of 5, 15, or 30 min or 1 h (Haskell Laboratory 1988a,b). The HF was diluted with either dry (<10% relative humidity) or humid (40-60% relative humidity) air. In this study, the acute toxicity of HF appeared to be related to relative humidity, and LC_{50} values were lower under the humid conditions. (LC_{50} values under dry and humid conditions, respectively, were 5 min, 14,640 ppm and 10,700 ppm; 15 min, 6,620 ppm and 2,470 ppm; 30 min, 2,890 and 1,020 ppm; and 1 h, 1,620 ppm and 540 ppm.) Subsequent work showed those values to be in error.

Because of the 3-fold difference in the two 1-h LC_{50} values at the high and low relative humidities, the above experiments were repeated using the same protocol but a different air sampling and collection device (Haskell Laboratory 1990). The results showed no effect of humidity on toxicity, but indicated difficulties with the sampling and analytical techniques in the earlier study. In this study, there was little difference between 1-h LC_{50} values at low (2,240 ppm) and high (2,340 ppm) relative humidities. Thus, the 1-h LC_{50} for head-only exposed rats was estimated at 2,300 ppm. Using probit analysis, LC_{01} values of 1,563 ppm (95% confidence limit, 1,004-1,781) and 1,827 ppm (95% confidence limit, 1,085-2,027) were reported for dry and humid air, respectively. Deaths occurred within the first 7 d post-exposure. Clinical signs were similar to those in the 1988 study and included signs of respiratory distress (labored breathing, lung noise, and/or gasping); ocular and nasal discharges, corneal opacity, necrotic lesions of the eyes, face, and ears; and severe weight loss among survivors during the 14-d recovery period. Most deaths occurred within 1-2 d of exposure, although a few deaths occurred up to 10 d post-exposure.

Dalbey et al. (1998a; see also Dalbey 1996) exposed groups of 10 or 20 female Sprague-Dawley rats to concentrations at 1,224 ppm or 2,039 ppm for 1 h; mortality was observed over a 14-d period. One death occurred in the 2,039-ppm group. In the same study, direct effects of HF on the trachea and lungs were studied using a mouth-breathing model in which groups of 10 or 20 rats were exposed via cannula to the trachea. This mouth-breathing model avoids the scrubbing effect of the nose. The durations of these experiments were 2, 10, or 60 min, the latter for comparison with mortality data in other studies. Each group of HF-exposed animals was compared with an identical group of sham-exposed controls. End points emphasized effects on the respiratory tract, the anticipated target site, but other organs were also evaluated. When the groups were composed of 20 rats, 10 rats were used for bronchoalveolar lavage (BAL), hematology, and serum chemistry. The remaining 10 were used for pulmonary function tests, histopathology, and organ weights. All animals were observed for clinical signs of toxicity immediately after exposure and before sacrifice on the following day. On the basis of preliminary results, sacrifice at 1 d after exposure provided data on the time of peak effects from HF. One group (exposed at 1,454 ppm for 10 min) was included to allow the authors to follow possible progression of lesions observed on the day after exposure; half of the animals were sacrificed at 3 wk after exposure, and the other half were sacrificed at 14 wk after exposure. Groups exposed at 3,847 ppm and 7,014 ppm for 10 min and 1,224 ppm and 2,039 ppm for 1 h were tested solely for mortality. Nose-breathing rats (plus orally cannulated animals) in groups exposed at 3,847 ppm and 7,014 ppm for 10 min were not sacrificed on the day after exposure, but were observed for 2 wk instead. Mortality in those groups was compared with published data on nose-breathing rats and allowed direct comparison of the mouth-breathing model with nose-breathing groups. Deaths were observed at the following concentrations and exposure times: 4,887 ppm and 8,621 ppm for 2 min (2/20 and 1/20 rats, respectively) and 1,764, 3,847, and 7,014 ppm for 10 min (1/20, 5/10, and 8/10 rats, respectively). No deaths occurred in orally cannulated rats inhaling 20 ppm or 48 ppm for 1 h. The data on orally cannulated rats are provided in Table 3-5 (above).

The primary sites of damage following acute exposures via cannula to the trachea appeared to be limited to the respiratory tract, particularly the trachea and bronchi. There was also evidence of effects in the lower lung resulting from orally cannulated exposures at the highest HF concentrations. In nose-breathing groups, the effects were generally limited to the nose; apparently the HF did not pass through the nose in sufficient amounts to affect the posterior sections of the respiratory tract. The ventral meatus was the site most affected in nose-breathing animals, followed by the nasoturbinates. The nasal septum was least affected. Necrosis and acute inflammation were noted in the nose; fibrinopurulent exudate was not. No significant lesions were noted in other organs examined microscopically, and no changes were observed in most of the other end points.

A definite exposure-related response was observed in orally cannulated rats exposed to HF for 2 min. At 8,621 ppm and 4,887 ppm, mortality was observed in 1/20 and 2/20 animals, respectively. Other evidence of serious toxicologic effects observed in animals in those groups included evidence of histopathologic damage in the lung (e.g., necrosis of the bronchial mucosa). Transient effects, including changes in BAL indices and flow at 25% forced vital capacity during forced exhalation were observed at 1,589 ppm. Histologic effects in the mid-trachea were also observed at that concentration. However, similar marginal effects in terms of incidence and severity were also observed in controls and may have resulted from cannulation. No deaths were observed in rats exposed at 593 ppm or 1,589 ppm for 2 min; no effects were observed at 593 ppm.

A concentration-response effect was also observed in orally cannulated rats exposed to HF for 10 min. Treatment-related mortality in one of 20 animals and serious toxicologic effects, including histopathologic damage in the lungs and trachea, were observed at 1,764 ppm. At 950 ppm, small increases in myeloperoxidase and polymorphonuclear leukocytes in the BAL were observed along with histologic changes in the trachea. These morphologic changes were marginal and were similar in incidence and severity to controls. No deaths were observed at 135, 271, 950, or 1,454 ppm for 10 min. No treatment-related effects were observed at 271 ppm.

In orally cannulated rats exposed to HF for 60 min, a minor increase in lung volume was observed across the upper part of the deflation pressure-volume curve at 48 ppm. No histologic changes were noted in the respiratory tract, and it was not clear that the change in the pressure-volume curve was an adverse effect. No effects were observed at 20 ppm for 60 min.

In the substudy on recovery, the effects of HF noted at the 1-d sacrifices were not observed in the animals at sacrifice at either 3 wk or 14 wk after exposure. The weights of the liver, spleen, and thymus were decreased at week 3 but not at week 14. These significant differences were associated with a significant decrease in the mean body weight compared with controls. The acute lesions essentially had resolved, and the tissues appeared to be repaired following the recovery period.

Two studies were conducted over longer exposure periods. In a range-finding study, groups of five male and five female Fischer-344 rats were exposed to measured concentrations of HF at 0 (air), 1, 10, 25, 65, or 100 ppm for 6 h/d, 5 d/wk for 14 d; survivors were sacrificed 2 d later (Placke et al. 1990). No deaths occurred in females inhaling 1 ppm or 10 ppm. Exposures at 25 ppm and above resulted in death in all females, with deaths beginning on the eighth, third, and second day of exposure at the 25-, 65-, and 100-ppm concentrations, respectively. No deaths occurred in males inhaling 1, 10, or 25 ppm; exposures at 65 ppm and 100 ppm resulted in death in all males, with deaths beginning on the third and second day at the 65-ppm and 100-ppm concentrations, respectively. No deaths occurred during the first day of exposure at any concentration. In the group exposed at 1 ppm, no effects other than a slight increase in lung-to-body weight ratio occurred. There were no clinical signs of toxicity in either gender at 1 and 10 ppm. There were no effects observed at 1 ppm except for a slight increase in absolute and relative lung weights in females and in absolute and relative heart weights in males. At 10 ppm and above, body weight and organ weight (liver, heart, kidney, and lungs) changes occurred in one or both genders. Clinical signs of nasal and ocular mucosal irritation occurred in the 25-, 65-, and 100-ppm groups. Dermal crust formation, ocular opacity, and tremors were also observed.

In a subchronic study (Placke and Griffin 1991), groups of 10 male and 10 female rats were exposed to concentrations of HF at 0.1, 1.0, or 10 ppm, administered as described above, for 91 d. Animals were observed for clinical signs, weighed, and subjected to hematology and blood chemistry examinations; tissues were examined microscopically. Five males and one female in the group exposed at 10 ppm died. Clinical signs included red-colored discharge from the eyes and nose, ruffled fur, alopecia, and hunched posture. Body weights were depressed, major organ weights were increased, and some blood parameters were changed compared with the control group. Dental malocclusions were observed in 11 animals. No deaths occurred at the two lower exposure concentrations.

Two groups of rats were exposed at 33 ppm (30 animals) or 8.6 ppm (15 animals) 6 h/d for a period of 5 wk (166 h) (Stokinger 1949). All rats died during the exposure at 33 ppm, whereas all rats survived the exposure period at the 8.6-ppm concentration. During exposure at the higher concentration, subcutaneous hemorrhages developed around the eyes and feet. Pathologic examinations at the end of the exposure period revealed moderate hemorrhage, edema, and capillary congestion in the lungs of 20 of 30 animals and renal-cortical degeneration and necrosis in 27 of 30 animals exposed at the higher exposure concentration.

3.1.3. Mice

Groups of 15 adult ICR mice were exposed to concentrations of HF ranging from 2,430 ppm to 11,010 ppm for 5 min (Higgins et al. 1972). Exposure concentrations were continuously monitored using specific ion electrodes. HF produced pulmonary edema of varying degrees of severity in most of the exposed mice. Pulmonary hemorrhage was a common finding in mice that died during or shortly after exposure at concentrations above the LC_{50} of 6,247 ppm. In exposures below the LC_{50}, delayed deaths occurred about 24 h after exposure; occasionally, deaths occurred 3-4 d later. No deaths occurred in mice exposed at 2,430 ppm for 5 min.

Groups of five male ICR mice were exposed to concentrations of HF at 500, 550, or 600 ppm for 1 h (MacEwen and Vernot 1970). Deaths occurred at all exposures; the LC_{50}, calculated by probit analysis, was 501 ppm (95% confidence limit, 355-705). In a similar study, groups of 10 female CF-1 mice were exposed at 351, 438, 505, 518, or 633 ppm for 1 h (MacEwen and Vernot 1974; Vernot et al. 1977). No deaths occurred at 351 ppm. The LC_{50}, calculated by probit analysis, was 456 ppm (95% confidence limit, 426-489).

Groups of 10 female ICR-derived mice were exposed to HF at 263, 278, 324, 381, or 458 ppm for 1 h (Wohlslagel et al. 1976). Animals were observed for toxic signs and mortality during a 14-d post-exposure period. Some animals that died following exposure or were sacrificed after the 14-d observation period were examined histologically. The 1-h LC_{50} was 342 ppm. Signs during the exposures included eye and mucous membrane irritation, respiratory distress, corneal opacity, and erythema of the exposed skin. Pathologic examinations of mice that died during or after exposure revealed pulmonary congestion and hemorrhage. No deaths occurred at 263 ppm.

In a repeated exposure study, two groups of mice were exposed at 33 ppm or 8.6 ppm 6 h/d for a period of 5 wk (166 h) (Stokinger 1949). All 18 mice died during the exposure at 33 ppm, whereas all mice survived the exposure period at the 8.6-ppm concentration. No pathologic examinations were undertaken.

3.1.4. Guinea Pigs

Young male guinea pigs of the Hartley strain were exposed in groups of 10 to various measured concentrations of HF for 15 min (Rosenholtz et al. 1963). The 15-min LC_{50} of 4,327 ppm was calculated following a 14-d observation period. At these concentrations, signs of irritation in the conjunctiva and nasal passages were observed and lasted 7 d post-exposure. They included reddened conjunctivae, pawing at the nose, marked lacrimation, nasal secretion, and sneezing. For animals surviving a week or more, respiratory distress, a body-weight loss of 25% during the first week, and general weakness were present. In the same study, a group of 10 young male guinea pigs was exposed at 1,377 ppm for 30 min. No deaths were reported; pathologic examinations were not performed.

Machle et al. (1934) exposed guinea pigs to concentrations ranging from 30 ppm to 9,760 ppm for exposure times of 5 min to 41 h. The data were summarized in a general manner by the authors and presented graphically (graph indicates approximately 100% mortality at 9,760 ppm for 5 min, at >4,000 ppm for 15 min, at >1,220 ppm for 2 h, and at >976 ppm for 3 h). According to the authors, a concentration at 1,220 ppm for 30 min did not produce death, but concentrations at >1,220 ppm to 1,830 ppm for as short a period of time as 5 min produced death in a significant number of animals. No deaths occurred in guinea pigs exposed at 122 ppm for 5 h.

In a follow-up study, Machle and Kitzmiller (1935) exposed three guinea pigs to a concentration of HF at 18.5 ppm for 6-7 h/d for 50 d—a total of 309 exposure h. After an initial weight gain, two guinea pigs lost weight and died during the exposures, one after 160 h of exposure and the other after approximately 250 h of exposure. Pathologic examinations of the two animals revealed the following lesions in one or both animals: pulmonary hemorrhage, inflammation and hyperplasia of the bronchial epithelium, congested and fatty liver with fibrotic changes, and renal tubular necrosis. The surviving animal was sacrificed 9 mo after the conclusion of the exposure. In that animal, the lungs showed hemorrhages, alveolar exudates, and alveolar wall thickening. The liver showed degeneration and necrosis.

3.1.5. Rabbits

Machle et al. (1934) exposed rabbits to concentrations ranging from 30 ppm to 9,760 ppm for exposure times of 5 min to 41 h. As noted above for guinea pigs, the data were summarized in a general manner by the authors and presented graphically (graph indicates approximately 100% mortality at 9,760 ppm for 5 min, at >4,000 ppm for 15 min, at >1,220 ppm for 2 h, and at >976 ppm for 3 h). According to the authors, a concentration at 1,220 for 30 min did not produce death, but concentrations at >1,220 ppm to 1,830 ppm for as short a period of time as 5 min produced deaths in a significant number of animals. No deaths occurred in rabbits exposed at 122 ppm for 5 h.

3.2. Nonlethal Toxicity

Data on effects following exposures at nonlethal concentrations of HF are available for the monkey, dog, rat, mouse, guinea pig, and rabbit. Data on single acute exposures are summarized in Table 3-6.

TABLE 3-6 Summary of Sublethal Effects of Hydrogen Fluoride Exposure in Laboratory Rats

Species	Concentration (ppm)	Exposure Time	Effect[a]	Reference
Dog	666	15 min	Moderate eye, nasal, and respiratory irritation; no changes in hematologic values	Rosenholtz et al. 1963
	460	15 min	Mild eye, nasal, and respiratory irritation	
Dog	243	1 h	Moderate eye, nasal, and respiratory irritation; no changes in hematologic values	Rosenholtz et al. 1963
	157	1 h	Mild eye, nasal, and respiratory irritation	
Rat	6,392	2 min	Inflammation, hemorrhage, necrosis of nasal epithelium (most animals); acute focal alveolitis of lung (1/20 animals)	Dalbey 1996; Dalbey et al. 1998b
Rat	2,432	5 min	Respiratory distress; severe eye and nasal irritation; weakness, sluggishness for 2 d	Rosenholtz et al. 1963
	1,438	5 min	Severe eye and nasal irritation	
	749	5 min	Moderate eye, nasal irritation	
Rat	7,014	10 min	Transient signs of ocular and nasal irritation, respiratory distress in 6/10 animals; severe rhinitis, mucus cell hyperplasia, mucosal necrosis in respiratory epithelium of nasal cavity; ocular damage in 2/10 animals	Dalbey 1996; Dalbey et al. 1998b
	3,847	10 min	Transient signs of ocular and nasal irritation, respiratory distress in 2/10 animals; mild to marked rhinitis, mucus cell hyperplasia, mucosal necrosis in respiratory epithelium of nasal cavity; ocular damage in 2/10 animals; respiratory depression	
	1,669	10 min	Inflammation, hemorrhage, necrosis of nasal area	

Rat	1,410	15 min	Respiratory distress; severe eye and nasal irritation; weakness, sluggishness for two days	Rosenholtz et al. 1963
	590	15 min	Moderate eye, nasal irritation	
	376	15 min	Mild eye, nasal irritation	
	307	15 min	Slight eye, nasal irritation	
Rat	1,377	30 min	Increase in activity; respiratory distress; severe eye and nasal irritation; no changes in body weight or organ/body weight ratios	Rosenholtz et al. 1963
Rat	100-1,000	30 min	Necrosis and inflammation restricted to the nasal region	Kusewitt et al. 1989; Stavert et al. 1991
	1,300	30 min	Immediate and persistent drop in ventilatory rate of 27%	
Rat	1,630[b] 1,910[c]	1 h	Respiratory epithelial inflammation and necrosis; no lung lesions	Haskell Laboratory 1990
Rat	1,224	1 h	Transient signs of ocular, nasal irritation, respiratory distress; moderate to severe rhinitis, mucus cell hyperplasia, mucosal necrosis of respiratory epithelium of nasal cavity; focal subacute alveolitis (2/10 animals); ocular damage (2/10 animals)	Dalbey 1996
Rat	489	1 h	Respiratory distress; severe eye and nasal irritation; weakness, sluggishness for two days	Rosenholtz et al. 1963
	291	1 h	Moderate eye, nasal irritation	
	126	1 h	Mild eye, nasal irritation	
	103	1 h	Occasional signs of eye and nasal irritation	

(Continued)

TABLE 3-6 Continued

Species	Concentration (ppm)	Exposure Time	Effect[a]	Reference
Rat	34	1 h	No nasal lesions, alveolitis (2/10 animals)	Dalbey 1996
Mouse	151	NA	Respiratory depression of 50%	TNO/RIVM 1996
Guinea pig	610, 964	5 min	Irritation, lung and organ lesions	Machle et al. 1934
Guinea pig	61	5 h	Mild irritation to respiratory tract	Machle et al. 1934
Guinea pig	54	6 h	Some liver and kidney damage	Machle et al. 1934
Rabbit	610, 964	5 min	Irritation, lung and organ lesions	Machle et al. 1934
Rabbit	1,247	15 min	Respiratory distress; severe eye and nasal irritation; weakness, sluggishness for two days; lung congestion in 1/5 rabbits	Rosenholtz et al. 1963
Rabbit	854	15 min	Moderate eye and nasal irritation; lung hemorrhage in 1/5 rabbits	Rosenholtz et al. 1963
Rabbit	61	5 h	Mild irritation to respiratory tract	Machle et al. 1934
Rabbit	54	6 h	Some liver and kidney damage	Machle et al. 1934

[a]Observed 24 h post-exposure (Kusewitt et al. 1989; Stavert et al. 1991; Dalbey 1996 [34 ppm for 1 h, 1,669 ppm for 10 min, 6,392 ppm for 2 min]), 7 d post-exposure (Higgins et al. 1972), 14 d post-exposure (MacEwen and Vernot 1970; Wohlslagel et al. 1976; Haskell Laboratory 1990; Dalbey 1996 [3,847 ppm and 7,014 ppm for 10 min, 1,224 ppm for 60 min]), up to 45 d post-exposure (Rosenholtz et al. 1963), or not specified (Machle et al. 1934).
[b]Low relative humidity.
[c]High relative humidity.
Abbreviation: NA, not available.

3.2.1. Nonhuman Primates

As noted in Section 3.1.1, no deaths were observed in four rhesus monkeys exposed to HF at 690 ppm for 1 h (MacEwen and Vernot 1970). In a longer-term study, two rhesus monkeys were exposed at 18.5 ppm for 6-7 h/d for 50 d—a total of 309 exposure h (Machle and Kitzmiller 1935). Except for an occasional cough during the first week of exposure, the animals appeared normal, and the concentration was considered tolerable and respirable. One monkey was sacrificed 8 mo post-exposure. The only prominent lesions were degenerative and inflammatory changes in the kidney.

3.2.2. Dogs

Groups of two mongrel dogs were exposed to HF at concentrations of 666 ppm or 460 ppm for 15 min or concentrations of 243 ppm or 157 ppm for 1 h and were observed for 14 d post-exposure (Rosenholtz et al. 1963). Those concentrations are approximately 25% and 12.5%, respectively, of the rat LC_{50}s for those respective time periods. During the exposures at 666 ppm or 243 ppm, the dogs showed signs of discomfort, including blinking, sneezing, and coughing. After removal from the exposure chambers, the dogs rubbed their noses and bodies on the grass. The cough persisted for 1-2 d and reappeared during the next 10 d only during periods of exercise. No skin lesions were noted. There were no changes in hematologic parameters (hematocrit and blood cell counts). Signs and effects were less severe at 460 ppm for 15 min and at 157 ppm for 1 h. Eye irritation was mild following exposure. Sneezing, rubbing of bodies on the ground, and a dry cough lasting 2 d were also observed following withdrawal. No gross lesions were noted, and no microscopic examinations were performed.

Two groups of dogs were exposed at 33 ppm (four dogs) or 8.6 ppm (five dogs) for 6 h/d for a period of 5 wk (166 h) (Stokinger 1949). No deaths occurred in either group. Pathologic examinations at the end of the exposure period revealed degenerative testicular changes (four of four animals), moderate hemorrhage and edema of the lungs (three of four animals), and ulceration of the scrotum (four of four animals) at the 33-ppm exposure concentration. At the lower exposure concentration, localized hemorrhagic areas were observed in the lungs of one of five animals. Clinical chemistry and hematology observations were unremarkable except for an increase in fibrinogen level at the higher exposure concentration.

3.2.3. Rats

Groups of 10 young Wistar-derived male rats were exposed to HF at various concentrations below the LC_{50} values for 5, 15, 30, or 60 min (Rosenholtz et al. 1963). Those concentrations were 2,432, 1,438, and 749 ppm for 5 min (approximately 50%, 25%, and 12.5% of the 5-min LC_{50}); 1,410, 590, 376, and 307 ppm for 15 min (approximately 50%, 25%, 12.5%, and 6% of the 15-min LC_{50}); 1,377 ppm for 30 min (68% of the 30-min LC_{50}); and 489, 291, 126, and 103 ppm for 60 min (approximately 50%, 25%, 12.5%, and 6% of the 60-min LC_{50}). Rats were observed for up to 45 d post-exposure. Clinical signs of toxicity included an increase in activity (at 68% of the LC_{50}) and conjunctival and nasal irritation. Symptoms diminished at lower concentrations. There were no significant body- or organ-weight changes. Tissues from 42 rats from across the exposure groups were examined microscopically. No lesions were present in the nasal passages, lungs, kidneys, liver, or bone marrow when concentrations were 50% or less of the LC_{50}.

Kusewitt et al. (1989) exposed Fischer-344 rats to concentrations of HF at 100-1,000 ppm for 30 min and sacrificed them 8 h and 24 h later. There was no mortality, and the lesions, necrosis and inflammation, were restricted to the nasal region. Histopathologic examinations and gravimetric measurements revealed no damage to the lungs. No further details were reported. In a related study by the same investigators, groups of five to eight male Fischer-344 rats were exposed at approximately 1,300 ppm for 30 min (Stavert et al. 1991). Ventilatory rates were measured during the exposure, and body weights and respiratory tract histology were investigated 24 h later. Rats exposed to HF experienced an immediate and persistent drop in ventilatory rate of 27%. A 10% reduction in body weight compared with nonexposed rats occurred by 24 h post-exposure. No changes in lung weights were observed. Changes in the nasal passages were limited to the anterior passages, with moderate to severe fibrinonecrotic rhinitis accompanied by large fibrin thrombi in the submucosa and hemorrhage. Lesions did not extend into the trachea. No deaths occurred by 24 h.

To evaluate respiratory tract effects, two groups of four rats were exposed for 1 h to HF at 1,630 ppm under conditions of low humidity or at 1,910 ppm under conditions high relative humidity. Groups of four were sacrificed at 1 d post-exposure and at 14 d post-exposure, and the respiratory tracts were examined histologically (Haskell Laboratory 1990). At 1 d after exposure, the examinations revealed that lesions were confined to the nose and were characterized by extensive acute necrosis and inflammation of the respiratory epithelium of the anterior nose with an inflammatory response in the submucosal tissue. No lesions were observed in the lower respiratory tract. At 14 d post-exposure, lesions were still present, but there was evidence of epithelial regeneration or repair in all rats.

A group of 20 Wistar rats was exposed to HF at 0.0016 mg/m^3 (0.002 ppm) for 5 h/d for 3 mo (Humiczewska et al. 1989). Compared with a group of control rats, exposed rats exhibited emphysemal changes in the lung involving enlarged alveoli and alveolar ducts and a narrowed interalveolar septum. Necrotic and hyperplastic areas were also noted. According to TNO/RIVM (1996), those changes were not clearly documented, and they are difficult to interpret.

3.2.4. Mice

The highest concentrations resulting in no mortality are reported in Table 3-4. In an unpublished ICI report cited in TNO/RIVM (1996), the respiratory rate of mice was halved (RD_{50}) at a concentration of 151 ppm (test range, 78-172 ppm). No further details were available.

3.2.5. Guinea pigs

Groups of three guinea pigs were exposed to concentrations of HF ranging from 30 ppm to 9,760 ppm for exposure durations of 5 min to 41 h (Machle et al. 1934). Respiratory tract irritation was observed at all concentrations and exposure periods. Symptoms included closed eyes, coughing and sneezing, mucoid conjunctival and nasal discharges, and slowing of the respiratory rate. Exposures above 2,440 ppm resulted in damage to the conjunctiva and nasal turbinates, pulmonary hemorrhages, and, in some cases, development of bronchopneumonia. Guinea pigs exposed at >610 ppm for 15 min or longer appeared weak and ill. Pathologic examinations revealed injury to the cornea and nasal mucous membranes; cardiac dilatation with congestion and myocardial injury; pulmonary hemorrhage, congestion, emphysema, edema, and bronchopneumonia; and hepatic, splenic, and renal congestion. Guinea pigs showed a tendency to delayed responses and deaths between the fifth and tenth week post-exposure. Concentrations below 122 ppm were tolerated for 5 h "without injury severe enough to produce death." Exposure at 61 ppm for approximately 5 h (exposure time read from graph) produced only mild irritation of the respiratory tract, and onset of symptoms was delayed compared with higher concentrations. Other post-exposure observation times were not clearly stated, but ranged from several hours to 15 wk. Additional data reported by the authors appear in Table 3-6. In the same study, exposure at 30 ppm for 6 h/d for approximately 5 d (41 h total) caused no deaths within a year following exposure, but lesions were present.

Guinea pigs were exposed to HF at either 33 ppm or 8.6 ppm for 6 h/d for a period of 5 wk (166 h) (Stokinger 1949). No deaths occurred with either exposure regime. No pathologic examinations were undertaken.

3.2.6. Rabbits

Groups of three rabbits were exposed to HF at 30-9,760 ppm for exposure durations of 5 min to 41 h (Machle et al. 1934). Concentrations below 122 ppm were tolerated for 5 h "without injury severe enough to produce death." Irritation of the eyes and respiratory tract and effects on other organs were the same as those described for the guinea pig above. Rabbits that did not contract bronchopneumonia returned to normal appearance and activity in a few days to a few weeks. In the same study, exposure at 30 ppm for 8 h/d for approximately 5 d (41 h total) caused no deaths; however, one rabbit examined 18 h after the last exposure had liver and kidney damage and fibrosing processes in the emphysematous lungs.

In a follow-up study, Machle and Kitzmiller (1935) exposed four rabbits to HF at 18.5 ppm for 6-7 h/d for 50 d—a total of 309 exposure hours. The animals gained weight throughout the exposure, although at a slower rate than a group of control rabbits. Pathologic examinations 7-8 mo post-exposure revealed the following lesions: leucocytic infiltration of the alveolar walls of the lungs; fatty changes in the liver; and degeneration, necrosis, and fibrosis of the kidneys. Two rabbits had acute lobular pneumonia. During metabolism studies, rabbits were exposed to concentrations ranging from 1.05 mg/L (1,283 ppm) for 1 h to 0.0152 mg/L (18.5 ppm) for 13 d (Machle and Scott 1935). Sacrifice occurred 9-15 mo later; no early deaths were reported.

Groups of five rabbits were exposed to concentrations of HF at 1,247 ppm or 854 ppm for 15 min (Rosenholtz et al. 1963). At the 1,247-ppm exposure, lacrimation, nasal discharge, pawing at the nose, reddened conjunctivae, and respiratory distress were observed, the latter lasting for a few hours after exposure. Those symptoms disappeared after 4 d. Signs were less severe at the lower concentration. Two exposed and two control rabbits were sacrificed and examined histologically. One rabbit exposed at the higher concentration (1,247 ppm) showed alveolar congestion at 14 d post-exposure, and one rabbit exposed at the lower concentration (854 ppm) showed severe intra-alveolar and intrabronchial hemorrhage when examined 2 d post-exposure.

Rabbits were exposed to HF at either 33 ppm or 8.6 ppm for 6 h/d for a period of 5 wk (166 h) (Stokinger 1949). No deaths occurred with either exposure regime. Slight pulmonary hemorrhage was observed in four of 10 rabbits at the higher exposure regime.

3.3. Developmental and Reproductive Toxicity

No studies were located addressing developmental or reproductive effects following inhalation exposure to HF. However, because effects on development and reproduction would be systemic (due to circulating fluoride), the effects of oral administration of fluoride are relevant. Those studies, reviewed by ATSDR (1993) and TNO/RIVM (1996), have conflicting results. Thus, the reproductive and developmental toxicity of HF cannot be fully assessed. However, studies indicate that there are no effects on animal reproduction and development when fluoride is administered at ≤400 ppm in drinking water.

Oral administration of sodium fluoride at 70 mg/kg for 5 d (Li et al. 1987) or at 75 ppm in drinking water for 21 wk (Dunipace et al. 1989) had no effect on spermatogenesis of $B6C3F_1$ mice. However, intraperitoneal injection of 8 mg/kg for five consecutive days (Pati and Buhnya 1987) and administration at 500 ppm or 1,000 ppm for up to 3 mo (DHHS 1991) resulted in abnormal spermatozoa in mice.

Sodium fluoride was administered in drinking water at 0, 10, 25, 100, 175, or 250 mg/L throughout gestation (Collins et al. 1995). At the highest dose level, maternal toxicity (reduced growth) and an increase in the number of fetuses with skeletal variations, but not the number of litters, was observed. No signs of retarded fetal development were observed, and the compound was not considered to have developmental toxicity.

Administration of sodium fluoride in drinking water at 0, 50, 150, or 300 ppm to pregnant rats during gestation days 6 through 15 or at 0, 100, 200, or 400 ppm to pregnant rabbits during gestation days 6 through 19 did not significantly affect the frequency of post-implantation loss, mean fetal weight/litter, or external, visceral, or skeletal malformations in either the rat or rabbit. Thus the NOAEL for developmental toxicity was >300 ppm (approximately 27 mg/kg/d) in the rat and >400 ppm (approximately 29 mg/kg/d) in the rabbit (Heindel et al. 1996).

3.4. Genotoxicity

Data on genotoxicity from inhalation exposures are limited. Voroshilin et al. (1975, as cited in ATSDR 1993) found hyperploidy in bone marrow cells of rats exposed at 1.0 mg/m^3 (1.22 ppm) for 6 h/d, 6 d/wk for 1 mo. The significance of hyperploidy is unknown. The same authors found no effects in C57B1 mice under the same conditions.

Other genotoxicity studies were conducted with sodium fluoride or potassium fluoride. Negative results were found in mutation studies with *Salmonella typhimurium* (with or without metabolic activation), and positive results were found in the mouse lymphoma (with and without activation), sister chromatid exchange (with and without activation), and chromosome aberration tests (without activation) (NTP 1990), generally at higher doses at which fluoride acts as a general "protein poison" (ATSDR 1993).

3.5. Chronic Toxicity and Carcinogenicity

No carcinogenicity studies using acute or longer-term inhalation exposures were located. Because inhaled HF would exert its systemic effects as fluoride ion, oral studies of fluoride administration may be relevant. A chronic oral carcinogenicity study in which sodium fluoride was administered to male and female rats and mice via drinking water resulted in equivocal evidence of bone cancer in male rats, but not in female rats or in mice of either gender (NTP 1990). The cancer was a rare bone osteosarcoma. Another chronic study (Maurer et al. 1990) found no evidence of cancers in male or female rats.

3.6. Summary

Sublethal and lethal inhalation data encompassing exposure times of 2 min to 6 h for six species of mammals were located. For 60-min exposures, only mild and occasional signs of ocular, nasal, or respiratory irritation were observed in the dog, at 157 ppm, and in the rat, at 103 and 126 ppm (Rosenholtz et al. 1963). The highest 60-min concentrations resulting in no deaths ranged from 263 ppm for the mouse to 1,087 ppm for the rat (Wohlslagel et al. 1976). No lung effects and only minor respiratory track effects were reported in orally cannulated rats exposed at 950 ppm for 10 min or 48 ppm for 1 h. One of 20 orally cannulated rats exposed at 1,764 ppm for 10 min died, whereas one of 10 nose-breathing rats exposed at 2,039 ppm for 1 h died. Severe irritant effects but no deaths occurred in nose-breathing rats exposed at 1,224 ppm for 1 h (Dalbey 1996; Dalbey et al. 1998a). The RD_{50} in mice was reported to be 151 ppm (TNO/RIVM 1996). Sixty-minute LC_{50} values ranged from a low of 342 ppm for the mouse (Wohlslagel et al. 1976) to a high of 2,300 ppm for the rat (Haskell Laboratory 1990). The lowest LC_{50} for the rat was 966 ppm (Vernot et al. 1977). Studies addressing developmental and reproductive effects, genotoxicity, and carcinogenicity generally used fluoride salts. The results of the two carcinogenicity studies were conflicting, but the shorter-term studies indicate that there are no exposure-related effects when fluoride is administered in drinking water.

4. SPECIAL CONSIDERATIONS

4.1. Metabolism/Disposition Considerations

4.1.1. Deposition

HF penetration in the lower respiratory tract may depend on concentration-exposure durations, because lower doses are effectively deposited only in the nasal passages, whereas very high concentrations reach the lungs. Species that are not obligate nose-breathers (e.g., humans) may also experience more severe lower respiratory tract effects, as would individuals undergoing exercise or physical exertion resulting in an increased ventilatory rate.

Morris and Smith (1982) studied the regional deposition of HF in the surgically isolated upper respiratory tract of male Long-Evans rats. Two endotracheal tubes were inserted into the upper tract, one for respiration and the other for collection of respired air. Known amounts of HF were drawn through the isolated upper respiratory tract while the animal respired HF-free air through an endotracheal tube. At fluoride concentrations ranging from 30 mg/m^3 to 176 mg/m^3 (HF at 38.5-226 ppm), >99.7% of the HF was deposited in the upper respiratory tract between the external nares and larynx; detectable levels did not penetrate to the trachea. A 6-h exposure of intact rats at 190 ppm resulted in 100% mortality by 3 h post-exposure but did not cause lung edema.

Because rats are obligatory nose-breathers while humans can breathe through both the nose and mouth, Stavert et al. (1991) studied the effects of HF inhalation via the nose and the mouth in rats. For compulsive mouth-breathing, male Fischer-344 rats were fitted with a mouthpiece and endotracheal tube, which extended to the middle of the trachea. Both nose-breathing and orally cannulated rats were exposed at approximately 1,300 ppm for 30 min. No deaths occurred in the nose-breathing rats, whereas 25% of orally cannulated rats died within 24 h of exposure. Effects in nose-breathing rats were limited to the anterior nasal passages; no changes were observed in the trachea. Observations were made at 24 h following exposure. Effects in orally cannulated rats occurred in the trachea (diffuse, severe fibrinonecrotic tracheitis and damaged tracheal rings), bronchi (focal areas of necrotizing bronchitis), and alveoli (scattered foci of polymorphonuclear leukocytes). Pulmonary function changes were observed, and lung weights were significantly elevated over values for nose-breathing rats. The results indicate that HF is effectively scrubbed in the nasal passages, and that species that are not obligate nose-breathers may be more sensitive to the effects of HF.

Likewise, in the study by Dalbey (1996; Dalbey et al. 1998a), orally cannulated rats were more susceptible to mortality than nose-breathing rats. No deaths occurred in nose-breathing rats exposed at 3,847 ppm or 7,014 ppm for 10 min, whereas mortality was high at the same concentrations (5/10 and 8/10, respectively) in orally cannulated rats.

Several studies indicated that there may be species differences in deposition of inhaled HF in the nasal passages and lungs, presumably due to differences in nasal scrubbing capacity. However, specific data reflecting differences in nasal scrubbing capacity could not be identified. The greater sensitivity of the rodents might largely be due to the rodents' higher respiratory rate. For example, all rats and mice exposed at 33 ppm for 5 wk died, whereas no deaths occurred in dogs, guinea pigs, and rabbits under the same exposure regime (Stokinger 1949). The greater sensitivity of the mouse compared with the rat, and the greater sensitivity of rodents compared with the monkey, as reflected by LC_{50} values (MacEwen and Vernot 1970; Wohlslagel et al. 1976), may be due to the higher respiratory rate of the mouse. Too few studies are available to predict regional deposition in humans, but Machle et al. (1934) noted irritation of only the larger airways from inhalation exposure at 32 ppm for several minutes.

4.1.2. Metabolism

HF is very soluble in water and is readily absorbed in the upper respiratory tract. The relatively low dissociation constant (3.5×10^{-4}) allows the non-ionized compound to penetrate the skin, respiratory system, or gastrointestinal tract and form a reservoir of fluoride ions that bind calcium and magnesium, forming insoluble salts (Bertolini 1992). Fluoride ion is absorbed into the bloodstream and is carried to all organs of the body in proportion to their vascularity and the concentration in the blood. Equilibrium across biological membranes is rapid (Perry et al. 1994). Significant deposition occurs in the bone, where the fluoride ion substitutes for the hydroxyl group of hydroxyapatite, the principal mineral component of bone. In humans, chronic exposure to elevated levels of fluoride or HF has produced osteofluorosis. Elimination is primarily through the kidneys.

Studies with human volunteers, workers exposed to HF in the workplace, and laboratory animals were located. Those studies, a few of which are summarized here, show that urinary excretion of F increases following exposure to HF and that chronic exposures may result in osteofluorosis. In five human subjects exposed to average air concentrations at 1.42-4.74 ppm 6 h/d for up to 50 d, fecal excretion of fluoride increased 4-fold from 0.102 mg/d to 0.423 mg/d (Largent 1960, 1961). Urinary excretion during the inhalation period ranged from 3.46 mg/d to 19.6 mg/d; pre-exposure urinary concentrations were not given. At an inhaled fluoride concentration of 4 ppm (measured from a combination of HF and silicon tetrafluoride) for 8 h, urinary excretion in two humans subjects increased from a pre-exposure range of 0.8-1.4 mg/d to an average of 9.1 mg during the day of exposure (Collings et al. 1951). Fluoride appeared in the urine within 2 h of exposure.

Mean urinary fluoride concentrations in aluminum smelter workers increased from a range of 1.23-2.45 mg/L pre-exposure to a range of 2.35-8.21 mg/L during a 7-d workshift (Dinman et al. 1976). Mean atmospheric concentrations of fluoride measured at different times and in different parts of the work area ranged from 0.73 mg/m^3 to 2.27 mg/m^3. At two other plants, average 8-h exposure concentrations ranged from 2.4 mg/m^3 to 6.0 mg/m^3 (2.9-7.3 ppm), and average fluoride concentrations in the urine ranged from 8.7 ppm to 9.8 ppm (Kaltreider et al. 1972). The urinary value for a control group was 0.7 ppm. Ninety-six percent of 79 chronically exposed employees who were X-rayed had developed fluorosis without physical impairment or overt clinical signs. On the basis of total fluoride levels, Derryberry et al. (1963) reported that exposure to a time-weighted average concentration of 3.38 mg/m^3 (range, 1.78-7.73 mg/m^3) was associated with increases in bone density in 17 of 74 workers. The average time of employment with fluoride exposure was 13.7 y. Exposure to a time-weighted concentration at 2.65 mg/m^3 did not result in a skeletal effect.

In a monitoring study, Kono et al. (1987) measured air and urinary concentrations of fluoride in 82 unexposed subjects and in 142 workers engaged in the manufacture of hydrofluoric acid in Japan. The air concentration for unexposed workers was 0 ppm, whereas the air concentrations in different areas of the manufacturing sites ranged from 0.3 ppm (16 workers) to 5.0 ppm (10 workers). There was a linear relationship between the mean values of urinary fluoride and the HF in the air.

Plasma fluoride concentrations significantly increased in male volunteers exposed to HF at 0.85-2.9 ppm or 3.0-6.3 ppm for 1 h (Lund et al. 1997). No increase was reported in a group exposed at 0.2-0.7 ppm for 1 h. In the higher exposure group, concentrations increased from approximately 8-17 ng/mL (baseline values for seven subjects) to 30-80 ng/mL at the end of the 1-h exposure. For several individuals in the highest exposure group, plasma fluoride remained elevated at 180 min after the start of exposure.

Rats exposed to airborne concentrations of HF (nose-only exposure) had increased concentrations of fluoride in their blood plasma and lungs (Morris and Smith 1983). There was a linear relationship between concentrations in air and plasma fluoride content. At an exposure concentration of 81 ppm, the blood and plasma ionic fluoride concentrations were 0.26 µg/ml and 0.19 µg/g, respectively. Mean concentrations in a control group were 0.037 µg/ml and 0.07 µg/g, respectively. The form of fluorine in both lung and plasma was primarily ionic, with a small amount being bound or nonexchangeable.

Plasma fluoride concentrations in guinea pigs exposed at 1.8, 3.7, 6.1, or 12.2 ppm were increased in a concentration-dependent manner (Dousset et al. 1986). After 84 h of constant exposure, plasma fluoride concentrations were 1.2, 1.4, 2, and 2.5 mg/L, respectively. Concentrations of F^- in the blood, urine, and bones of guinea pigs exposed to atmospheres containing 0.6-29 mg/m^3 (0.7-35 ppm) for one day increased with increasing HF concentrations (Bourbon et al. 1984).

Following short-term HF exposures in rabbits, the bulk of stored fluorine was found in the bone, whereas following long-term exposures, storage also occurred in the teeth (Machle and Scott 1935). For example, 15 mo after a single 1-h exposure at 1,281 ppm, the fluorine content in bone was 99.4 mg/100 g. The value in a group of unexposed controls was 14.25 mg/100 g of bone. Storage of as much as 10 times the amount normally found in bone caused changes normally associated with fluorosis. After 16 d of exposure at 33 ppm for 6 h/d, the fluoride content of the teeth and femurs of rats increased by a factor of 10-30 (Stokinger 1949).

4.2. Mechanism of Toxicity

The available studies indicate that HF is a severe irritant to the skin, eyes, and respiratory tract, particularly the anterior nasal passages where, depending on species and concentration, it appears to be effectively scrubbed from the inhaled air. Effective deposition in the anterior nasal passages may be attributed to the high aqueous solubility and reactivity of HF. Penetration into the lungs results in pulmonary hemorrhage and edema and may result in death. Cardiac arrhythmias have been seen in humans following accidental dermal and inhalation exposure. Cardiac arrhythmias are the result of hypocalcemia- and hypomagnesemia-induced acidosis following dermal fluoride uptake. It is not known whether inhalation exposure alone would cause this effect (ATSDR 1993). Although renal and hepatic changes have been observed in animal studies, serious systemic effects are unlikely to occur at a level below what would cause serious respiratory effects. In the studies summarized in Tables 3-4 and 3-6, the tissues of the respiratory tract sustain the impact of an acute exposure. Therefore, the concentration of HF in the inhaled air, and not the absorbed dose, is the primary determinant of effects in acute exposure scenarios.

4.3. Structure-Activity Relationships

Although the AEGL values for HF are based on empirical toxicity data, it is important to consider the relative toxicities of HF and other structurally similar chemicals. The compounds most closely related to HF are the other hydrogen halides, HCl and HBr. It might be anticipated that some relationships exist in this chemical class between structure and respective toxicities in animals and humans. However, because of the differences in size and electron configuration of the various halogen atoms, substantial differences exist with respect to their chemical and physical properties, which in turn are responsible for their toxicologic properties (atomic weights of fluorine, chlorine, and bromine are 19, 35.5, and 80, respectively). That is particularly true in the case of acute toxic effects resulting from inhalation exposure.

For example, HCl has a considerably higher ionization constant than HF, and is therefore classified as a stronger acid. Consequently, higher concentrations of proton-donor hydronium ions are generated from HCl in aqueous solutions under the same conditions. The protons readily react with cells and tissues resulting in HCl's irritant and corrosive properties. On the other hand, the fluoride ion from dissociated HF is a strong nucleophile, or Lewis base, that is highly reactive with various organic and inorganic electrophiles, which are biologically important substances, also resulting in irritation and tissue damage.

In addition to these differences in chemical properties, differences in water solubility may be a significant factor in acute inhalation toxicity. HF and HBr are characterized as infinitely and freely soluble in water, respectively, and the solubility of HCl, although high, is lower at 67 g/100 g of water at 30°C (Budavari et al. 1996). Thus, it is likely that HF is more effectively scrubbed than HCl in the nasal cavity, resulting in less penetration to the lungs and less severe toxicity there. The effectiveness of the scrubbing mechanism is demonstrated in a study that addressed the acute toxicities of HF, HCl, and HBr and the deposition (scrubbing) of those chemicals in the nasal passages. Stavert et al. (1991) exposed male Fischer-344 rats to each of the hydrogen halides at 1,300 ppm for 30 min and assessed damage to the respiratory tract 24 h after the exposure. The nasal cavity was divided into four regions, which were examined microscopically. For all three hydrogen halides, tissue injury was confined to the nasal cavity. Tissue injury in the nasal cavity was similar following exposures to HF and HCl and involved moderate to severe fibrinonecrotic rhinitis in nasal region 1 (most anterior region). For HF and HCl, the lesions extended into region 2, but regions 3 and 4 were essentially normal in appearance, as was the trachea. Nasal cavity lesions following exposure to HBr were limited to region 1 and were similar in extent to those produced by HF and HCl, showing that all three chemicals are well scrubbed. No lung or tracheal injury was evident for any of the chemicals, although accumulations of inflammatory cells and exudates in the trachea and lungs following the exposure to HCl indicated that it may not be as well scrubbed in the nasal passages as HF and HBr. However, that possibility is modified by the authors' observation of lower minute volumes in the HF- and HBr-exposed rats, so that greater amounts of HCl were breathed. Morris and Smith (1982) also showed that at concentrations up to 226 ppm, >99.7% of inspired HF may be scrubbed in the upper respiratory tract of the rat.

In a series of experiments with HF and HCl that used guinea pigs and rabbits as the test species, Machle and coworkers (Machle and Kitzmiller 1935; Machle et al. 1934, 1942) concluded that the acute irritant effects of HF and HCl were similar, but the systemic effects of HF were more severe, presumably because chloride ion is a normal electrolyte in the body, and fluoride ion is not. However, the conclusions involving systemic effects followed repeated exposures.

Aside from lethality studies, no clear evidence is available to establish the relative toxicities of HF and HCl. At concentrations ranging from 100 ppm to 1,000 ppm for 30 min, Kusewitt et al. (1989) reported epithelial and submucosal necrosis, accumulation of inflammatory cells and exudates, and extravasation of erythrocytes in the nasal region of rats exposed to HF, HCl, or HBr. The severity of injury increased with increasing concentration, and the relative toxicities of the hydrogen halides were reported as HF > HCl > HBr. However, in a later study by the same authors, Stavert et al. (1991) reported no difference in the toxicities of HF, HCl, or HBr to the nasal regions or the lung in nose-breathing or mouth-breathing rats, respectively, at 1,300 ppm for 30 min.

At the high concentrations necessary to cause lethality during exposure durations of 5 min to 1 h, HF is approximately twice as toxic to the rat (1.8- to 2.2-fold) as HCl (Table 3-7). The relationship is similar for the mouse within that time period (2.2- to 3.2-fold). A later study on HF by MacEwen and Vernot (1974) resulted in slightly lower 1-h LC_{50} values of 456 ppm and 966 ppm for the mouse and rat, respectively, but the approximately 2-fold difference between HF and HCl remained for both species. However, when considering lethal concentrations of respiratory irritants such as HCl, the mouse "may not be an appropriate model for extrapolation to humans," because "mice appear to be much more susceptible to the lethal effects of HCl than other rodents or baboons. To some extent, this increased susceptibility may be due to less effective scrubbing of HCl in the upper respiratory tract" (NRC 1991). Quantitative data for HBr were limited to one study, but that study also showed that HF was more toxic than either HCl or HBr.

It is important to note that the relative toxicities for HF, HCl, and HBr shown in Table 3-7 are for exposure durations ranging from 5 min to 1 h. On the basis of empirical lethality (LC_{50}) data in rats, rabbits, and guinea pigs, the exposure time-LC_{50} relationship for HF using the equation $C^n \times t = k$ results in an n value of 2. That compares to a value of $n = 1$ empirically derived from rat and mouse lethality data for HCl. Hence, although HF is more toxic than HCl at the higher concentrations and shorter exposure durations, the rate of decrease in the LC_{50} threshold is less (i.e., less slope in the curve derived from $C^n \times t = k$) for HF than HCl. As a result, the LC_{50}

values, and therefore the lethal toxicities of HF and HCl, are comparable at 4 h and 8 h. This shift in relative lethal toxicity across time also is reflected in the AEGL-3 values developed for HF and HCl.

Considering the greater water solubility of HF compared with HCl, it is possible that the more effective scrubbing of HF in the nasal passages is responsible for the apparent decrease in the relative toxicities of HF and HCl at lower concentrations associated with longer exposure durations. Conversely, the greater toxicity of HF at higher concentrations associated with shorter exposure durations might be due to saturation of the scrubbing mechanism and higher concentrations in the lower respiratory system.

4.4. Other Relevant Information

4.4.1. Susceptible Populations

Individuals with asthma might respond to exposure to HF with increased bronchial responsiveness. No information on the relative susceptibility of asthmatic and normal individuals to HF was located.

Individuals under stress, such as those involved in emergency situations and individuals engaged in physical activity, will experience greater HF deposition and pulmonary irritation than individuals at rest. Furthermore, individuals who breathe through their mouths would be at greater risk. The exercise incorporated into the protocol of the Lund et al. (1997, 1999) study takes into account the increased physical activity in emergency situations.

4.4.2. Species Differences

Lethal concentrations were investigated in rats, mice, and guinea pigs. Results of studies in different species by the same investigators indicate that the rat is more sensitive than the guinea pig (Rosenholtz et al. 1963), and the mouse is more sensitive than the rat (Wohlslagel et al. 1976). Similar 60-min LC_{50} values for the rat were found by Wohlslagel et al. (1976), Rosenholtz et al. (1963), and MacEwen and Vernot (1970)—1,395, 1,307, and 1,276 ppm, respectively. The monkey, with its lower respiratory rate, is less sensitive to HF than the mouse or rat (by factors of 1.4 and 3.5, respectively [MacEwen and Vernot 1970, 1971]). Data presented in this document clearly show that the rank order of susceptibility is mice > rats > rhesus monkeys.

TABLE 3-7 Relative Toxicities of HF, HCl, and HBr as Expressed by LC_{50} Values (ppm)

Species	Exposure Duration	HF	HCl	HBr	Reference
Rat	5 min	18,200	41,000		Higgins et al. 1972
Mouse		6,247	13,750		
Rat	30 min	2,042	4,700		Rosenholtz et al. 1963 (HF); MacEwen and Vernot 1972 (HCl)
Mouse			2,644		
Rat	1 h	1,395	3,124		Wohlslagel et al. 1976[a]
Mouse		342	1,108		
Rat	1 h	1,278		2,858	MacEwen and Vernot 1972[a]
Mouse		501		814	

[a]The data of Wohlslagel et al. (1976) and MacEwen and Vernot (1972) were generated in the same laboratory. Therefore, the values for HCl can be compared with those for HF and HBr in the following row.

When comparing species differences in sensitivity, there are four studies that indicate that mice are 2 to 4 times more sensitive to the toxic effects of HF than are rats. Three of those studies are summarized in Table 3-7, above. The 5-min LC_{50} value for the mouse was 6,247 ppm compared with 18,200 for the rat (Higgins et al. 1972). In the second study, the 60-min LC_{50} for the mouse was 342 ppm compared with 1,395 ppm for the rat (Wohlslagel et al. 1976). In the third study, the 1-h LC_{50} values for the mouse and rat were 501 ppm and 1,276 ppm, respectively (MacEwen and Vernot 1970). In the fourth study, the 1-h LC_{50} of the rat was 966 ppm, whereas that of the mouse was 456 ppm (Vernot et al. 1977). When mouse and rat lethality values are compared across all studies, the species difference appears greater than 3- to 4-fold, but that greater difference is probably attributable to the different analytical techniques among laboratories. Monkeys have substantially lower respiratory rates than rodents and are therefore less susceptible to pulmonary irritation from inhaled HF.

In other studies, the monkey was less sensitive than the rat (1-h LC_{50} values of 1,774 ppm and 1,276 ppm, respectively). The rat was more sensitive than the guinea pig (15-min LC_{50} values of 2,689 ppm and 4,327 ppm, respectively) (Rosenholtz et al. 1963). Responses of rabbits and guinea pigs to sublethal and lethal exposures appeared to be similar, except that guinea pigs were more likely than rabbits to suffer delayed deaths. Some differences between the two species were noted in the severity and incidence of organ lesions (Machle et al. 1934). In another study, rats and mice suffered 100% mortalities at a concentration of 33 ppm for 6 h/d for 5 wk, whereas no mortalities occurred in guinea pigs, rabbits, and dogs (Stokinger 1949). Rhesus monkeys survived exposure at 18.5 ppm for 6-7 h/d for 50 d, with lesions confined to the kidneys, whereas two of three guinea pigs died during the same exposure regimen (Machle and Kitzmiller 1935). Lesions were present in the lungs and organs of guinea pigs. Rabbits survived the same exposure regimen, but suffered lung and organ lesions. Differences in nasal scrubbing capacity among the species could not be ascertained from these data.

Because most rodents are obligatory nose-breathers, whereas humans may be mouth-breathers, especially during exercise, Stavert et al. (1991) and Dalbey (1996) studied the effects of HF inhalation via the nose and mouth in rats. HF was delivered directly to the trachea by cannulation. In both studies, concentrations that produced effects confined to the nasal passages in nose-breathing rats resulted in serious lower respiratory tract effects or deaths in orally cannulated rats. These results indicate that the site of injury and resultant toxicologic effects will differ with oral and nasal breathing, the former mode resulting in more severe responses under similar exposure situations. Thus, species that can breathe through their mouth may be more sensitive to the effects of HF than are those who are obligate nose-breathers.

4.4.3. Concentration-Exposure Duration Relationship

When data are lacking for desired exposure times, time-scaling can be executed based on the relationship between acute toxicity (concentration) and exposure duration for a common end point. Time-scaling data were available for the rat (Rosenholtz et al. 1963). Using 5-, 15-, 30-, and 60-min LC_{50} values from Rosenholtz et al. (1963), Alexeeff et al. (1993) showed that the association between concentration and exposure duration for HF can be described as $C^2 \times t = k$ (where C = concentration, t = time, and k is a constant). The least-squares linear curve fit of the graph, log time vs log LC_{50}, resulted in the equation $y = 7.69 - 1.89x$. The slope of the line, 1.89 (rounded to 2), is the value of the exponent n. The graph showing this relationship is in Appendix A. ten Berge et al. (1986) found the same relationship between concentration and time using the data of Machle et al. (1934) (i.e., $C^2 \times t = k$).

The above time-scaling relationship holds true for exposure durations between 5 and 60 min. Because HF is well scrubbed in the nasal passages at "low concentrations," time-scaling to longer exposure durations results in lower values that are inconsistent with the experimental data.

5. DATA ANALYSIS FOR AEGL-1

5.1. Summary of Human Data Relevant to AEGL-1

The studies of Largent (1960, 1961), Machle et al. (1934), and Lund et al. (1997, 1999) provide information on both concentrations and exposure times. These studies with healthy human volunteers, including subjects who were characterized as atopic, indicated that a concentration of HF at 1.42 ppm was a NOAEL for irritation, although average concentrations of 2.59-4.74 ppm could be tolerated for 6 h daily for a number of days with only slight irritation of the eyes, skin, and nasal passages (Largent 1960, 1961). Concentrations at ≤2 ppm (Largent 1960, 1961) and ≤3 ppm (Lund et al. 1997, 1999) were characterized as only slightly irritating; there was a slight lung inflammatory response in exercising adults in the study by Lund et al. (1999). Moreover, excursions up to 7.9 ppm and 8.1 ppm (Largent 1960, 1961) could be tolerated by two subjects, apparently without increased irritancy. Workers have been chronically exposed to HF at an average concentration of 3.6 ppm (Derryberry et al. 1963), ranges of concentrations from 0.3 ppm to 5.0 ppm (Kono et al. 1987) and from 0.2 ppm to 5 ppm (Abramson et al. 1989), and intermittently at 14 ppm to 27 ppm (Machle and Evans 1940). Those workers were for the most part asymptomatic (Derryberry et al. 1963; Machle and Evans 1940), although the chronic exposures in the study of Derryberry et al. (1963) were associated with increases in bone density (osteofluorosis).

5.2. Summary of Animal Data Relevant to AEGL-1

Studies involving mild irritation effects in animals were summarized in Table 3-6. For 60-min exposures, mild and occasional signs of eye, nose, or respiratory irritation were observed in the dog at 157 ppm and in the rat at 103 and 126 ppm (Rosenholtz et al. 1963). For guinea pigs and rabbits, a concentration at 61 ppm for 5 h resulted in only mild irritation to the respiratory tract, but histologic studies were not performed at that exposure concentration.

5.3. Derivation of AEGL-1

Because human data are available, they should be used to derive the AEGL-1. The basis for the AEGL-1 is 3 ppm, a higher exposure concentration in a range of concentrations evaluated by Lund et al. (1997, 1999). The exposure duration was 1 h. This exposure level can be considered a subthreshold for lung inflammation, because there were no increases in markers of lung inflammation, which include neutrophils, eosinophils, protein, and methyl histamine. There were no changes in lung function (FVC and FEV_1), and no to minor symptoms of irritation at this concentration. The subjects were healthy exercising adults, but two of them had increased immune parameters. Compared with healthy adults, individuals with asthma may experience bronchial constriction at lower concentrations of HF.

For HF, the use of an intraspecies uncertainty factor (UF) of 3 for the Lund et al. (1997, 1999) data is reasonable. There was no evidence of effects on respiratory parameters in healthy adults at concentrations up to 7.8 ppm (Lund et al. 1995) or 8.1 ppm (Largent 1960, 1961) or in healthy but atopic individuals at concentrations up to 6.3 ppm (Lund et al. 1997, 1999). Application of an intraspecies UF of 3 to data sets that included atopic individuals results in a 1-h AEGL-1 value of 1 ppm. This value is lower than the concentrations in the Lund et al. (1995, 1997, 1999) and Largent (1960, 1961) studies by factors of 6-8 and is considered sufficiently low to protect asthmatic individuals. Because irritant properties would not change greatly between the 10-min and 1-h time frames, the 10- and 30-min values were set equal to the 1-h value. The resulting AEGL-1 values are listed in Table 3-8, and the calculations are contained in Appendix B.

The Largent et al. (1960, 1961) study provides support data for the safety of the longer-term AEGL-1 values. In the Largent study, concentrations at ≤2 ppm for 6 h/d were reported as only slightly irritating. This concentration-time relationship for an acute exposure can be considered conservative, because the exposure was tolerated repeatedly for up to 50 d without increased irritancy. In addition, industrial exposures at approximately 4 ppm have been experienced without effects.

Alexeeff et al. (1993) used a benchmark dose approach to estimate an exposure level that would protect the public from any irritation from routine HF emissions. Their approach employed a log-probit extrapolation of concentration-response data to the 95% lower confidence limit on the toxic concentration producing a benchmark dose of 1% response called a practical threshold. Species-specific and chemical-specific adjustment factors were applied to develop exposure levels applicable to the general public.

TABLE 3-8 AEGL-1 Values for Hydrogen Fluoride (ppm [mg/m^3])

10 min	30 min	1 h	4 h	8 h
1.0 (0.8)	1.0 (0.8)	1.0 (0.8)	1.0 (0.8)	1.0 (0.8)

The 1-h value was calculated to be 0.7 ppm. That value is similar to the 1.0-ppm concentration derived for the 1-h AEGL-1. Alexeeff et al. (1993) also calculated a 1-h value of 2 ppm, which they defined as the concentration that would protect against severe irritation from a once-in-a-lifetime release.

6. DATA ANALYSIS FOR AEGL-2

6.1. Summary of Human Data Relevant to AEGL-2

Studies with human volunteers indicate that because of irritation, concentrations above 32 ppm cannot be tolerated for more than a few minutes (Machle et al. 1934). The highest concentration that two men could voluntarily tolerate for more than 1 min was 122 ppm. Longer-term exposures were at lower average concentrations, 2.59-4.74 ppm, but excursions up to 8 ppm were tolerated daily for 25-50 d without more than the described mild irritant effects (Largent 1960, 1961). Furthermore, long-term exposure to combined HF gas and fluoride-containing dust at 3.6 ppm (Derryberry et al. 1963), 8 h of exposure to HF and silicon tetrafluoride at approximately 4 ppm (Collings et al. 1951), and intermittent long-term exposure to combined HF and calcium fluoride dust at 14-27 ppm (Machle and Evans 1940) did not result in clinical or respiratory effects in workers. According to Alarie (1981), the mouse RD_{50} of 151 ppm (TNO/RIVM 1996) would be "intolerable to humans" and would result in tissue damage, but 0.1 times the RD_{50} (15 ppm) would result in only "some sensory irritation" and could be tolerated for hours to days. This latter irritant effect is less than the irritation considered relevant to the development of AEGL-2 values.

6.2. Summary of Animal Data Relevant to AEGL-2

Five animal studies (Machle et al. 1934; Rosenholtz et al. 1963; Higgins et al. 1972; Wohlslagel et al. 1976; Dalbey 1996) provide information on sublethal effects from acute exposures. The 5-min exposure periods of

Higgins et al. (1972) are short and, aside from a mention of pulmonary edema, details of effects were not given. One-hour exposures at 157 ppm (dog) and 103 ppm (rat) produced mild, nondisabling effects or only occasional signs of ocular and nasal irritation (Rosenholtz et al. 1963). One-hour exposures at 243 ppm (dog) and 291 ppm (rat) produced moderate ocular, nasal, and respiratory irritation (Rosenholtz et al. 1963). For the guinea pig and rabbit, a concentration at 61 ppm for 5 h resulted in only mild irritation of the respiratory tract, but histologic studies revealed that exposures at 54 ppm for 6 h and 30 ppm for 41 h resulted in liver and kidney lesions (Machle et al. 1934).

The Dalbey (1996; Dalbey et al. 1998a) data are relevant to the development of short-term AEGL-2 values. In those studies, 10-min exposures of orally cannulated rats (a conservative model for the human breathing pattern under irritant conditions, because HF would not be scrubbed by the mouth and upper respiratory tract) resulted in a NOAEL of 950 ppm for the AEGL-2 definition (the next highest exposure, 1,764 ppm, resulted in death in one of 20 rats). At 950 ppm, small increases in myeloperoxidase and polymorphonuclear leukocytes in the BAL were observed along with histologic changes in the trachea. These morphologic changes were marginal and were similar in incidence and severity to controls. It should be noted that in nose-breathing rats, a concentration at 1,669 ppm was well scrubbed by the nasal passages, and lesions were confined to the nasal passages (Dalbey et al. 1998b).

6.3. Derivation of AEGL-2

6.3.1. Derivation of 10-Min AEGL-2

Animal data for short-term exposures and human data for long-term exposures were available for consideration in calculating the AEGL-2 values. Because 10-min data were available in orally cannulated rats, they were used to derive the 10-min AEGL-2. Mortality occurred at the 10-min, 1,764-ppm exposure in cannulated rats, so the lower value of 950 ppm was chosen as the threshold for serious effects for the 10-min AEGL-2. However, no serious effects occurred at that exposure even though HF was delivered directly to the trachea. A total UF of 10 is reasonable because the end point chosen is a mild response that does not approach the definition of an AEGL-2 effect. In addition, the delivery of HF in orally cannulated rats eliminates scrubbing by the nose. The dose to the lungs is greater than in naturally breathing rodents. Therefore, this route of exposure coupled with the mild effect is already an inherently conservative exposure from which to derive the 10-min AEGL-2. Using this estimated threshold, a total UF of 10 (3 each for inter- and intraspecies differences) was applied resulting in a 10-min AEGL-2 of 95 ppm. The 10-min AEGL-2 is clearly below the serious injury category of data from tests in monkeys, rats, dogs, mice, guinea pigs, and rabbits, as shown in Figure 3-1. Therefore, a total UF of 10 applied to the orally cannulated rat data should be protective of susceptible populations. Figure 3-1 is a plot of all of the data on HF from 3 min to 7 h by category of response.

6.3.2. Derivation of 30-Min to 8-H AEGL-2s

The 1-h exposure of dogs at 243 ppm, which resulted in signs of irritation and discomfort including blinking, sneezing, and coughing, was chosen as the basis for the longer-term AEGL-2 values. That value is one-fourth of the rat LC_{50} in the same study. Rats exposed to a similar concentration (291 ppm) developed moderate eye and nasal irritation. The next higher concentration (489 ppm for 1 h) resulted in respiratory distress and severe eye and nasal irritation in the rat, signs more severe than those ascribed to AEGL-2. The moderate eye and nasal irritation observed in dogs at 243 ppm was considered the threshold for impaired ability to escape. That 1-h value was divided by a total UF of 10 (3 for intraspecies differences and 3 for interspecies differences). The values were scaled across time using $C^2 \times t = k$. The resulting time-scaled 8-h AEGL-2 value of 8.6 ppm is inconsistent with the human data in the study of Largent (1960, 1961) in which humans inhaling 8.1 ppm intermittently suffered no greater effects than slight irritation. Therefore, the 8-h AEGL-2 was set equal to the 4-h AEGL-2. It should be noted that the resulting 30-min AEGL-2 of 34 ppm is similar to the 32-ppm concentration that was tolerated for only several minutes by human subjects in the Machle et al. (1934) study. The AEGL-2 values are listed in Table 3-9, and the calculations are contained in Appendix B.

A total UF of 10 is reasonable and sufficient. If a total UF of 30 were used, the predicted 6-h AEGL-2 level would be 3.3 ppm. However, human subjects exposed intermittently at 8 ppm for 6 h over a 10-50 d period experienced only slight irritation (Largent 1960, 1961). Even a susceptible person should not experience a disabling effect at that concentration. The derived 6-h value using a total UF of 10 is 9.9 ppm, which is in the range

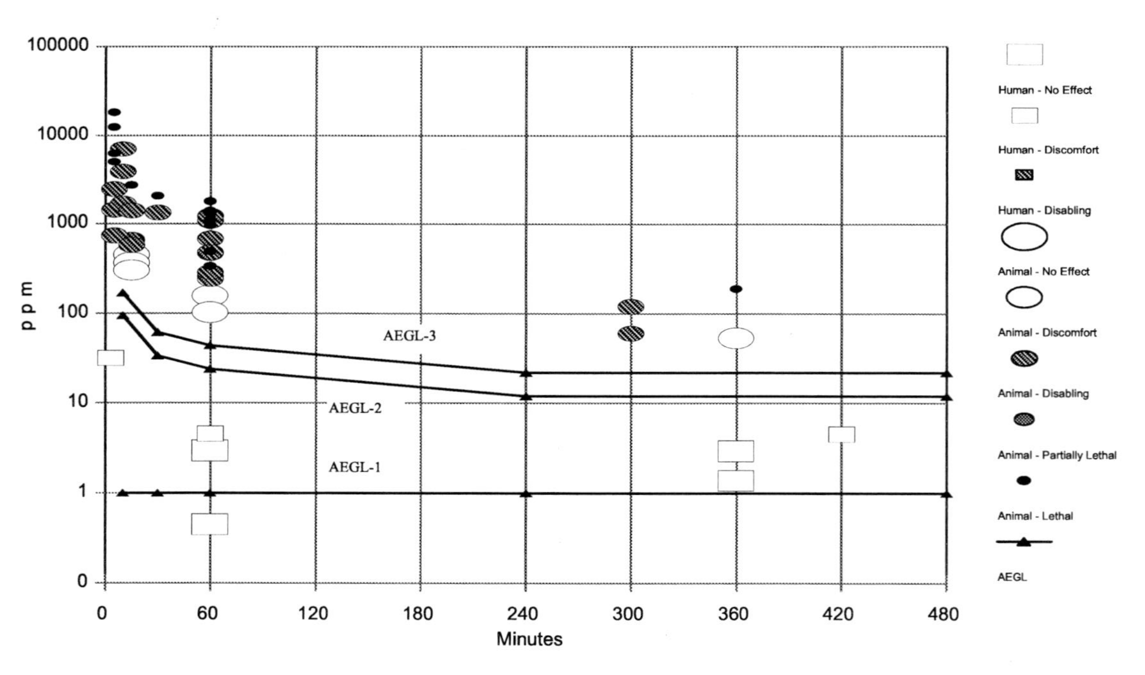

FIGURE 3-1 Toxicity data and AEGL values for hydrogen fluoride. Toxicity data include both human and animal studies.

TABLE 3-9 AEGL-2 Values for Hydrogen Fluoride (ppm [mg/m^3])

10 min	30 min	1 h	4 h	8 h
95 (78)	34 (28)	24 (20)	12 (9.8)	12 (9.8)

that a healthy person can tolerate with only minor irritation upon repeated exposure. Susceptible individuals, including asthmatic patients, should not be incapacitated at that concentration. Figure 3-1 is a plot of all of the data on HF from 3 min to 7 h by category of response. The AEGL-2 values are clearly below the serious injury category of data from tests in monkeys, rats, dogs, mice, guinea pigs, and rabbits.

According to Alarie (1981), 0.1 times the RD_{50} of 151 ppm for the mouse (TNO/RIVM 1996) can be tolerated for hours by humans with some irritation. The resulting concentration of 15 ppm is only slightly higher than both the 4- and 8-h values in Table 3-9.

7. DATA ANALYSIS FOR AEGL-3

7.1. Summary of Human Data Relevant to AEGL-3

Human data were insufficient for derivation of an AEGL-3. The highest nonlethal concentration that human volunteers were exposed to, 122 ppm for 1 min, produced marked irritation (Machle et al. 1934). That is a lesser effect than the AEGL-3 definition. Scaling to the longer time periods would result in lower values than the AEGL-2. Fatalities of humans from HF exposure have been reported, but exposures were via both dermal and inhalation routes, and exposure concentrations were unknown. Exposures during accidents were usually estimated.

7.2. Summary of Animal Data Relevant to AEGL-3

Lethality data are summarized in Table 3-4, and sublethal data are summarized in Table 3-6. The lowest lethal concentrations were reported for the mouse. Wohlslagel et al. (1976) reported a 1-h LC_{50} of 342 ppm and a 1-h value of 263 ppm that resulted in no deaths, which the data indicate is a threshold for lethality. However, the LC_{01}, calculated by probit analysis, was 200 ppm, below the LC_0. The data also indicate that mice are approximately 3 times more sensitive to HF than rats.

The Dalbey (1996; Dalbey et al. 1998a) data are relevant to the development of short-term AEGL-3 values. In those studies, 10-min exposures to HF in orally cannulated rats (a potentially realistic model for the human breathing pattern under irritant conditions) resulted in serious effects including lethality at 1,764 ppm (1/20) and caused local irritation but no serious effects at 950 ppm. The NOAEL was 271 ppm.

7.3. Derivation of AEGL-3

7.3.1. Derivation of 10-Min AEGL-3

The concentration causing death in one of 20 orally cannulated rats, 1,764 ppm, was chosen as the lethal threshold for the 10-min AEGL-3 (Dalbey et al. 1998a). Although 1/20 is higher than the usual threshold for the AEGL-3, the oral cannulation model is conservative compared with normal nose-breathing, because it bypasses nasal scrubbing and maximizes the pulmonary dose. A total UF of 10 is reasonable under those conditions. Figure 3-1 shows that the 10-min AEGL-3 value is clearly below levels that cause death in tests in monkeys, rats, dogs, mice, guinea pigs, and rabbits. The use of a total UF of 30 would drive the 10-min AEGL-3 to 59 ppm, which is below the 10-min AEGL-2. Nose-breathing rats exposed at 1,669 ppm for 10 min had lesions confined to the nasal passages (Dalbey et al. 1998b). In addition, exposure of monkeys at 690 ppm for 1 h (MacEwen and Vernot 1970) did not result in any deaths. That value extrapolated to 10 min with an $n = 2$ and a total UF of 10 would predict no deaths if monkeys were exposed at 169 ppm for 10 min, essentially the same value as the 10-min AEGL-3 of 170 ppm. Therefore, a total UF of 10 applied to the orally cannulated rat data should be quite protective.

7.3.2. Derivation of 30-Min to 8-H AEGL-3s

The AEGL-3 values for the threshold for life-threatening health effects or death for the longer time periods were based on the 60-min value of 263 ppm, which was the highest nonlethal concentration for the mouse reported by Wohlslagel et al. (1976). UFs and MFs are applied to account for interspecies and intraspecies variability. On the basis of LC_{50} data from studies in which the susceptibility of both the rat and mouse were evaluated, the mouse was found to be approximately three times more sensitive than the rat to the effects of HF (Table 3-7). Because of the greater sensitivity of the mouse, an interspecies UF of 1 was applied; an intraspecies UF of 3 was applied to account for differences in human susceptibility. In addition, a MF of 2 was applied, because the highest nonlethal value of 263 ppm was close to the 60-min LC_{50} of 342 ppm. The 30-min and 4-h values were calculated based on the $C^2 \times t = k$ relationship. Because the time-scaled 8-h value of 15 ppm is inconsistent with the animal data (e.g., two rhesus monkeys survived a 50-d exposure to 18.5 ppm with no effects other than kidney lesions [Machle and Kitzmiller 1935]), the 8-h AEGL-3 value was set equal to the 4-h value. Calculations are contained in Appendix B, and results are listed in Table 3-10.

A total factor of 6 is reasonable and sufficient. If a total factor of 20 were used (3 each for inter- and intraspecies uncertainties and 2 as an MF), the predicted 6-h AEGL-3 would be 5.4 ppm. However, human subjects exposed intermittently at 8 ppm for 6 h over a 25-50 d period experienced only slight irritation (Largent 1960, 1961). Even a susceptible person should not experience a life-threatening effect at that concentration. In addition, the use of a total factor of 20 would drive the AEGL-3 values below the AEGL-2 values. Therefore, a combined factor of 6 is reasonable. The AEGL-3 values are clearly below the levels that cause death in monkeys, rats, dogs, mice, guinea pigs, and rabbits (Figure 3-1).

The database for HF is extensive, and many other studies with laboratory animals support the AEGL-3 values. Concentrations resulting in severe effects but no deaths in several animal species were divided by 10 (an interspecies UF of 3 to account for the fact that these species may be less sensitive than mice, and an intraspecies UF of 3 to account for differences in human susceptibility). When scaled across time, the 1-h values ranged between 69 ppm and 163 ppm, which are above the 1-h AEGL-3 of 44 ppm. In addition, the RD_{50} for mice of 151 ppm is not far below the 10-min AEGL-3 of 170 ppm. According to Alarie (1981), the RD_{50} could be tolerated for hours by humans but would result in tissue damage.

TABLE 3-10 AEGL-3 Values for Hydrogen Fluoride (ppm [mg/m^3])

10 min	30 min	1 h	4 h	8 h
170 (139)	62 (51)	44 (36)	22 (18)	22 (18)

8. SUMMARY OF AEGLs

8.1. AEGL Values and Toxicity End Points

In summary, the AEGL values were derived in the following manner. The AEGL-1 was based on the study of Lund et al. (1997, 1999) in which sensitive biochemical markers of pulmonary inflammation were noted in human volunteers at a mean concentration of 3 ppm. The exposures were for 1 h. In a supporting study, human volunteers could tolerate exposure at a concentration of 2 ppm for 6 h/d with only mild irritation of the eyes, skin, and upper respiratory tract (Largent 1960, 1961). The 3-ppm concentration from the Lund et al. (1997, 1999) study was divided by an intraspecies UF of 3 to protect sensitive individuals. Because of the 6-h exposure duration of the supporting study, and because adaptation to mild sensory irritation occurs, the 1 ppm value was applied to all AEGL-1 exposure durations.

The 10-min AEGL-2 value was based on the 10-min NOAEL of 950 ppm for serious lung effects in orally cannulated rats; that value was considered the threshold for serious effects. A combined UF of 10—3 for interspecies variability (HF is a primary irritant, LC_{50} values differ by a factor of 2-4 between the mouse and rat, and the irritation end point is appropriate for human health risk assessment) and 3 for intraspecies variability.

The 30-min and 1-, 4- and 8-h AEGL-2 values were based on a study in which dogs exposed at 243 ppm for 1 h showed signs of irritation including blinking, sneezing, and coughing. The 1-h value of 243 ppm was divided by a total UF of 10—3 for intraspecies (the dog is a sensitive species for sensory irritation) and 3 for intraspecies. The values were scaled across time using $C^2 \times t = k$. Based on the study of Largent (1960, 1961) in which human subjects intermittently exposed at 8.1 ppm for up to 50 d found the exposure only slightly irritating, the calculated time-scaled 8-h value of 8.6 ppm was considered too low. Therefore, the 8-h value was set equal to the 4-h value.

The 10-min AEGL-3 was based on the 10-min lethal threshold in orally cannulated rats, 1,764 ppm. That value was rounded down to 1,700 ppm and divided by a combined UF of 10—3 for interspecies differences (LC_{50} values differ by a factor of 2-4 between the mouse and rat) and 3 for intraspecies differences.

The other AEGL-3s were derived from the highest reported nonlethal 1-h value for the most sensitive animal species, the mouse. That value, 263 ppm, was divided by UFs of 1 for interspecies variability (the mouse was

TABLE 3-11 Summary of AEGL Values (ppm [mg/m³])

Classification	Exposure Duration 10 min	30 min	1 h	4 h	8 h
AEGL-1 (Nondisabling)	1.0 (0.8)	1.0 (0.8)	1.0 (0.8)	1.0 (0.8)	1.0 (0.8)
AEGL-2 (Disabling)	95 (78)	34 (28)	24 (20)	12 (9.8)	12 (9.8)
AEGL-3 (Lethal)	170 (139)	62 (51)	44 (36)	22 (18)	22 (18)

approximately 3 times more sensitive than rats) and 3 for intraspecies variability and by an MF of 2 to account for the fact the highest nonlethal value was close to the LC_{50}. That value was scaled to the other time periods using $C^2 \times t = k$. On the basis of repeated exposures in animal studies and the well-known scrubbing capacity of the mammalian nose at low concentrations, the 8-h AEGL-3 was set equal to the 4-h value. Values are summarized in Table 3-11, above.

8.2. Comparisons with Other Standards and Criteria

Standards and guidelines developed by other agencies are listed in Table 3-12. Those values are for both daily exposures in the workplace and emergency situations. The ERPGs are defined similarly to the AEGLs but are usually for only the 1-h exposure duration. Although based on older data than those used for AEGL development, the 1-h ERPG levels for the respective categories are similar to the 1-h AEGLs. In 1999, 10-min ERPGs were developed for HF. For the three respective levels, they were 2, 50, and 170 ppm. The ERPG-1 and ERPG-3 values are similar to the respective 10-min AEGLs.

The NIOSH IDLH (NIOSH 2002) of 30 ppm, applicable to a 30-min time period prior to donning suitable protective equipment or evacuation, is similar to the 30-min AEGL-2 of 34 ppm. The IDLH of 30 ppm is based on eye, nose, and lung irritation seen in studies with humans (Largent 1961) and animals (Machle et al. 1934).

Other standards and guidelines, applicable to an 8-h work day or peak excursions during an 8-h work day (3 ppm), are higher than the AEGL-1 (1 ppm), indicating the conservative end point chosen for the AEGL-1 and the fact that the AEGL-1 is applicable to the general population.

TABLE 3-12 Extant Standards and Guidelines for Hydrogen Fluoride (ppm)

	Exposure Duration				
Guideline	0 min	30 min	1 h	4 h	8 h
AEGL-1	1.0	1.0	1.0	1.0	1.0
AEGL-2	95	34	24	12	12
AEGL-3	170	62	44	22	22
ERPG-1 (AIHA)[a]	2		2		
ERPG-2 (AIHA)	50		20		
ERPG-3 (AIHA)	170		50		
CEEL-1 (CIC)[b]					1.5
CEEL-2 (CIC)					7
CEEL-3 (CIC)					50
IDLH (NIOSH)[c]		30			
REL-TWA (NIOSH)[d]					3[e]
REL-STEL (NIOSH)[f]					6[e]
PEL-TWA (OSHA)[g]					3[e]
TLV-Ceiling (ACGIH)[h]					3[e]
MAK (Germany)[i]					3
MAC-Peak Category (The Netherlands)[j]					3.3

[a]ERPG (emergency response planning guidelines, American Industrial Hygiene Association) (AIHA 2002). The ERPG-1 is the maximum airborne concentration below which it is believed nearly all individuals could be exposed for up to 1 h without experiencing any symptoms other than mild, transient adverse health effects or without perceiving a clearly defined objectionable odor. The ERPG-2 is the maximum airborne concentration below which it is believed nearly all individuals could be exposed for up to 1 h without experiencing or developing irreversible or other serious health effects or symptoms that could impair an individual's ability to take protective action. The ERPG-3 is the maximum airborne concentration below which it is believed nearly all individuals could be exposed for up to 1 h without experiencing or developing life-threatening health effects.
[b]CEEL (community emergency exposure levels) (Clement International Corp., unpublished material). CEELs are concentrations that cause adverse health effects

(exposure durations are unspecified). CEELs I, II, and III are designated alert, evacuation, and death levels, respectively. The CEELs developed by Clement International Corp. are recommended values.
[c]IDLH (immediately dangerous to life and health, National Institute of Occupational Safety and Health) (NIOSH 2002). IDLH represents the maximum concentration from which one could escape within 30 min without any escape-impairing symptoms or any irreversible health effects. [d]NIOSH REL-TWA (recommended exposure limit–time-weighted average, National Institute of Occupational Safety and Health) (NIOSH 2002). The REL-TWA is defined analogous to the ACGIH-TLV-TWA (i.e., the time-weighted average concentration for a normal 8-h work day and a 40-h work week, to which nearly all workers may be repeatedly exposed, day after day, without adverse effect).
[e]As fluorine.
[f]NIOSH REL-STEL (recommended exposure limits–short-term exposure limit) (NIOSH 2002). The REL-STEL is defined analogous to the ACGIH TLV-STEL (i.e., it is defined as a 15-min TWA exposure that should not be exceeded at any time during the work day even if the 8-h TWA is within the TLV-TWA). Exposures above the TLV-TWA up to the STEL should not be longer than 15 min and should not occur more than 4 times per day. There should be at least 60 min between successive exposures in this range.
[g]OSHA PEL-TWA (permissible exposure limits–time-weighted average, Occupational Health and Safety Administration) (NIOSH 2002). The PEL-TWA is defined analogous to the ACGIH TLV-TWA, but is for exposures of no more than 10 h/d, 40 h/wk.
[h]ACGIH TLV-ceiling (Threshold Limit Value–ceiling, American Conference of Governmental Industrial Hygienists) (ACGIH 2002). This is the concentration that should not be exceeded during any part of the working day.
[i]MAK (Maximale Arbeitsplatzkonzentration [Maximum Workplace Concentration]) (Deutsch Forschungsgemeinschaft [German Research Association] 2000). The MAK is defined analogous to the ACGIH TLV-TWA. The peak category for HF is 1.
[j]MAC (Maximaal Aanvaaarde Concentratie [Maximal Accepted Concentration-peak category]) (SDU Uitgevers [under the auspices of the Ministry of Social Affairs and Employment], The Hague, The Netherlands 2000). The MAC is defined analogous to the ACGIH TLV-ceiling.

8.3. Data Adequacy and Research Needs

The database for human exposures is relatively good—both short-term and long-term studies with human volunteers are available. The well-conducted study of Lund et al. (1997, 1999) examined lung dynamics as well as biomarkers of exposure. In addition, some data from industrial exposures were located, although few exposures were to HF alone. Data from animal studies used six species of mammals and encompassed a wide range of exposure concentrations and a range of exposure durations, including daily exposures. The data from human studies was adequate to derive or support the AEGL-1 and AEGL-2 concentrations, and the database for animal studies was adequate to derive the AEGL-2 and AEGL-3 concentrations. Extrapolation to longer exposure times for the animals studies showed that, based on the data of Largent (1960, 1961), the values were too low. Therefore, the 8-h AEGL-2 and AEGL-3 values were set equal to the respective 4-h values.

Sampling and analytical methods used in the human and animal studies conducted in the 1960s and 1970s were not as sensitive as those perfected by the late 1980s and 1990s and may have under- or overestimated concentrations. An improved sampling/analytical methodology developed by Haskell Laboratory (1990) indicates that HF may have collected on glassware in the exposure apparatus. That factor would indicate that exposure concentrations in the early studies may have been underestimated. However, the studies of Lund et al. were very recent and used a more sensitive analytical method.

9. REFERENCES

Abramson, M.J., J.H. Wlodarczyk, N.A. Saunders, and M.J. Hensley. 1989. Does aluminum smelting cause lung disease? Am. Rev. Resp. Dis. 139:1042-1057.

ACGIH (American Conference of Governmental Industrial Hygienists). 1996. Threshold Limit Values (TLVs) for Chemical and Physical Agents and Biological Exposure Indices (BEIs). Cincinnati, OH: ACGIH.

ACGIH (American Conference of Governmental Industrial Hygienists). 1997. Threshold Limit Values (TLVs) for Chemical and Physical Agents and Biological Exposure Indices (BEIs). Cincinnati, OH: ACGIH.

ACGIH (American Conference of Governmental Industrial Hygienists). 2002. Documentation of the Threshold Limit Values and Biological Exposure Indices: Fluorine, 6th Ed. Cincinnati, OH: ACGIH.

AIHA (American Industrial Hygiene Association). 2000. Emergency Response Planning Guidelines and Workplace Exposure Level Guides. Fairfax, VA: AIHA.

Alarie, Y. 1981. Dose-response analysis in animal studies: Prediction of human responses. Environ. Health Perspect. 42:9-13.

ATSDR (Agency for Toxic Substances and Disease Registry). 1993. Toxicological Profile for Fluorides, Hydrogen Fluoride, and Fluorine (F). TOP-91/17. Agency for Toxic Substances and Disease Registry, Public Health Service, Washington, DC.

Alexeeff, G.V., D.C. Lewis, and N.L. Roughly. 1993. Estimation of potential health effects from acute exposure to hydrogen fluoride using a benchmark dose approach. Risk Anal. 13:63-69.

Amoore, J.E., and E. Hautala. 1983. Odor as an aid to chemical safety: Odor thresholds compared with Threshold Limit Values and volatilities for 214 industrial chemicals in air and water dilution. J. Appl. Toxicol. 3:272-290.

Bertolini, J.C. 1992. Hydrofluoric acid: A review of toxicity. J. Emerg. Med. 10:163-168.

Bourbon, P., C. Rioufol, and P. Levy. 1984. Relationships between blood, urine and bone F levels in guinea pig after short exposures To HF. Fluoride 17:124-131.

Braun, J., H. Stoss, and A. Zober. 1984. Intoxication following inhalation of hydrogen fluoride. Arch. Toxicol. 56:50-54.

Budavari, S., M.J. O'Neil, A. Smith, and P.E. Heckelman, eds. 1996. The Merck Index, 12th Ed. Rahway, NJ: Merck & Co., Inc.

Chan, K.-M., W.P Svancarek, and M. Creer. 1987. Fatality due to acute hydrofluoric acid exposure. Clin. Toxicol. 25:333-339.

Chela, A., R. Reig, P. Sanz, E. Huguet, and J. Corbella. 1989. Death due to hydrofluoric acid. Am. J. Forensic Med. Pathol. 10:47-48.

Clement International Corporation. [Undated]. Monograph for hydrogen fluoride/hydrofluoric acid. Prepared for U.S. Environmental Protection Agency, Office of Pollution, Prevention and Toxics, Washington, DC.

Collins, T.F.X., R.L. Sprando, M.E. Shackleford, T.N. Black, et al. 1995. Developmental toxicity of sodium fluoride in rats. Food Chem. Toxicol. 33:951-960.

Collings, G.H., R.B.L. Fleming, and R. May. 1951. Absorption and excretion of inhaled fluorides. AMA Arch. Ind. Hyg. Occup. Med. 4:585-590.

Dalbey, W. 1996. Evaluation of the Toxicity of Hydrogen Fluoride at Short Exposure Times. Petroleum Environmental Research Forum Project 92-09, performed at Stonybrook Laboratories Inc., Pennington, NJ.

Dalbey, W., B. Dunn, R. Bannister, W. Daughtrey, C. Kirwin, F. Reitman, A. Steiner, and J. Bruce. 1998a. Acute effects of 10-minute exposure to hydrogen fluoride in rats and derivation of a short-term exposure limit for humans. Regul. Toxicol. Pharmacol. 27:207-216.

Dalbey, W., B. Dunn, R. Bannister, W. Daughtrey, C. Kirwin, F. Reitman, M. Wells, and J. Bruce. 1998b. Short-term exposures of rats to airborne hydrogen fluoride. Toxicol. Environ. Health 55(Part A):241-275.

Dayal, H.H., M. Brodwick, R. Morris, T. Baranowski, N. Trieff, J.A. Harrison, J.R. Lisse, and G.A.S. Ansari. 1992. A community-based epidemiologic study of health sequelae of exposure to hydrofluoric acid. Ann. Epidemol. 2:213-330.

Derryberry, O.M., M.D. Bartholomew, and R.B.L. Fleming. 1963. Fluoride exposure and worker health. Arch. Environ. Health 6:65-73.

DHHS (U.S. Department of Health and Human Services). 1991. Review of Fluorides: Benefits and Risks. Report of the Ad Hoc Committee on Fluoride of the Committee to Coordinate Environmental Health and Related Programs. U.S. Department of Health and Human Services, Washington, DC.

Dinman, B.D., W.J. Bovard, T.B. Bonney, et al. 1976. Excretion of fluoride during a seven-day workweek. J. Occup. Med. 18:14-16.

DiPasquale, L.C., and H.V. Davis. 1971. Acute toxicity of brief exposures to hydrogen fluoride, hydrogen chloride, nitrogen dioxide, and hydrogen cyanide singly and in combination with carbon monoxide. AMRL-TR-71-120, AD-751-442. National Technical Information Service, Springfield, VA.

Dousset, J.C., C. Rioufel, P. Bourbon, P. Levy, and R. Feliste. 1986. Effects of inhaled HF on cholesterol metabolism in guinea pigs. Fluoride 19:71-77.

Dunipace, A.J., W. Zhang, T.W. Noblitt, et al. 1989. Genotoxic evaluation of chronic fluoride exposure: Micronucleus and sperm morphology studies. J. Dent. Res. 68:1525-1528.

EPA (U.S. Environmental Protection Agency). 1993. Hydrogen Fluoride Study: Final Report, Report to Congress, Section 112(n)(6), Clean Air Act as amended. EPA 550-R-93-001. U.S. Environmental protection Agency, Washington, DC.

German Research Association (Deutsche Forschungsgemeinschaft). 1999. List of MAK and BAT Values, 1999. Commission for the Investigation of Health Hazards of Chemical Compounds in the Work Area, Report No. 35. Weinheim, Federal Republic of Germany: Wiley VCH.

Haskell Laboratory. 1988a. Preliminary Report on the Concentration-Time-Response Relationships for Hydrogen Fluoride with Attachments and Cover Letter dated 062188. EPA/OTS; Doc #FYI-OTS-0788-0607 (Same as R. Valentine. 1988. Preliminary Report on the Concentration-Time-Response Relationships for Hydrogen Fluoride. Du Pont HLR 281-88).

Haskell Laboratory. 1988b. Test results of acute inhalation studies with anhydrous hydrogen fluoride with cover letter dated 031688. TSCATS/303700, EPA/OTS; Doc #FYI-OTS-0388-0607.

Haskell Laboratory. 1990. Acute inhalation toxicity of hydrogen fluoride in rats (final report) with attachments and cover letter dated 082390. EPA/OTS; Doc #FYI-OTS-0890-0607.

Heindel, J.J., H.K. Bates, C.J. Price, M.C. Mark, C.B. Myers, and B.A. Schwetz. 1996. Developmental toxicity evaluation of sodium fluoride administered to rats and rabbits in drinking water. Fundam. Appl. Toxicol. 30:162-177.

Higgins, E.A., V. Fiorca, A.A. Thomas, and H.V. Davis. 1972. Acute toxicity of brief exposures to HF, HCl, NO_2, and HCN with and without CO. Fire Technol. 8:120-130.

Himes, J.E. 1989. Occupational medicine in Oklahoma: Hydrofluoric acid dangers. J. Okla. State Med. Assoc. 82:567-569.

Humiczewska, M., W. Kuzna, and A. Put. 1989. Studies on the toxicology of fluorine compounds. I. Histological and histochemical investigations on the liver, heart, lungs, and stomach of rats exposed to hydrogen fluoride [in Polish]. Folia Biol. (Krakow) 37:181-186.

Kaltreider, N.L., M.J. Elder, L.V. Crally, and M.O. Colwell. 1972. Health survey of aluminum workers with special reference to fluoride exposure. J. Occup. Med. 14:531-541.

Kleinfeld, M. 1965. Acute pulmonary edema of chemical origin. Arch. Environ. Health 10:942-946.

Kono, K., Y. Yoshida, H. Yamagata, M. Watanabe, Y. Shibuya, and K. Doi. 1987. Urinary fluoride monitoring of industrial hydrofluoric acid exposure. Environ. Res. 42:415-420.

Kusewitt, D.F., D.M. Stavert, G. Ripple, T. Mundie, and B.E. Lehnert. 1989. Relative acute toxicities in the respiratory tract of inhaled hydrogen fluoride, hydrogen bromide, and hydrogen chloride. Toxicologist 9:36.

Largent, E.J. 1960. The metabolism of fluorides in man. AMA Arch. Ind. Health 21:318-323.

Largent, E.J. 1961. Fluorosis: The Health Aspects of Fluorine Compounds. Columbus, OH: Ohio State University Press. Pp. 34-39, 43-48.

Lee, D.C., J.F. Wiley, and J.W. Snyder. 1993. Treatment of inhalation exposure to hydrofluoric acid with nebulized calcium gluconate. J. Occup. Med. 35:470.

Li, Y.M., A.J. Dunipace, and G.K. Stookey. 1987. Effects of fluoride on the mouse sperm morphology test. J. Dent. Res. 66:1509-1511.

Lund, K., M. Refsnes, P. Sostrand, P. Schwarze, J. Boe, and J. Kongerud. 1995. Inflammatory cells increase in bronchoalveolar lavage fluid following hydrogen fluoride exposure. Am. J. Respir. Crit. Care Med. 151:A259.

Lund, K., J. Ekstrand, J. Boe, P. Sostrand, and J. Kongerud. 1997. Exposure to hydrogen fluoride: An experimental study in humans of concentrations of fluoride in plasma, symptoms, and lung function. Occup. Environ. Med. 54:32-37.

Lund, K., M. Refsnes, T. Sandstrom, P. Sostrand, P. Schwarze, J. Boe, and J. Kongerud. 1999. Increased CD3 positive cells in bronchoalveolar lavage fluid after hydrogen fluoride inhalation. Scand. J. Work Environ. Health 25:326-334.

MacEwen, J.D., and E.H. Vernot. 1970. Toxic Hazards Research Unit Annual Technical Report: 1970. AMRL-TR-70-77, AD 714694. Aerospace Medical Research Laboratory, Wright-Patterson AFB, Ohio.

MacEwen, J.D., and E.H. Vernot. 1971. Toxic Hazards Research Unit Annual Technical Report: 1971. AMRL-TR-71-83. Aerospace Medical Research Laboratory, Wright-Patterson AFB, Ohio.

MacEwen, J.D., and E.H. Vernot. 1972. Toxic Hazards Research Unit Annual Technical Report: 1972. AMRL-TR-72-62, AD-755 358. Aerospace Medical Research Laboratory, Wright-Patterson AFB, Ohio.

MacEwen, J.D., and E.H. Vernot. 1974. Toxic Hazards Research Unit Annual Technical Report: 1974. AMRL-TR-74-78. Aerospace Medical Research Laboratory, Wright-Patterson AFB, Ohio.

Machle, W., and E.E. Evans. 1940. Exposure to fluorine in industry. J. Ind. Hyg. Toxicol. 22:213-217.

Machle, W., and K. Kitzmiller. 1935. The effects of the inhalation of hydrogen fluoride. II. The response following exposure to low concentrations. J. Ind. Hyg. 17:223-229.

Machle, W., and E.W. Scott. 1935. The effects of the inhalation of hydrogen fluoride. III. Fluorine storage following exposure to sub-lethal concentrations. J. Indust. Hyg. 17:230-240.

Machle, W., F. Thamann, K. Kitzmiller, and J. Cholak. 1934. The effects of the inhalation of hydrogen fluoride. I. The response following exposure to high concentrations. J. Ind. Hyg. 16:129-145.

Machle, W., K.V. Kitzmiller, E.W. Scott, and J.F. Treon. 1942. The effect of the inhalation of hydrogen chloride. J. Ind. Hyg. 24:222-225.

Maurer, J.K., M.C. Chang, B.G. Boysen, et al. 1990. 2-Year carcinogenicity study of sodium fluoride in rats. J. Natl. Cancer Inst. 82:118-126.

Mayer, T.G., and P.L. Gross. 1985. Fatal systemic fluorosis due to hydrofluoric acid burns. Ann. Emerg. Med. 14:149-153.

Ministry of Social Affairs and Employment (SDU Uitgevers). 2000. Nationale MAC (Maximum Allowable Concentration) List, 2000. Ministry of Social Affairs and Employment, The Hague, The Netherlands.

Morris, J.B., and F.A. Smith. 1982. Regional deposition and absorption of inhaled hydrogen fluoride in the rat. Toxicol. Appl. Pharmacol. 62:81-89.

Morris, J.B. and F.A. Smith. 1983. Identification of two forms of fluorine in tissues of rats inhaling hydrogen fluoride. Toxicol. Appl. Pharmacol. 71:383-390.

NIOSH (National Institute for Occupational Safety and Health). 1976. Criteria for a Recommended Standard: Occupational Exposure to Hydrogen Fluoride. DHHS (NIOSH) Publication 76-143. U.S. Department of Health and Human Services, Washington, DC.

NIOSH (National Institute for Occupational Safety and Health). 1995. Hydrogen Fluoride. Documentation for Immediately Dangerous to Life or Health Concentrations (IDLHs). U.S. Department of Health and Human Services, National Institute for Occupational Safety and Health, Cincinnati, OH [Online]. Available: http://www.cdc.gov/niosh/idlh/7664393.html [2002].

NRC (National Research Council). 1989. Biologic Markers in Pulmonary Toxicology. Washington, DC: National Academy Press.

NRC (National Research Council). 1991. Permissible Exposure Levels and Emergency Exposure Guidance Levels for Selected Airborne Contaminants. Washington, DC: National Academy Press.

NRC (National Research Council). 1993. Guidelines for Developing Community Emergency Exposure Levels for Hazardous Substances. Washington, DC: National Academy Press.

NRC (National Research Council). 2001. Standing Operating Procedures for Developing Acute Exposure Guideline Levels for Hazardous Chemicals. Washington, DC: National Academy Press.

NTP (National Toxicology Program). 1990. Technical Report on the Toxicology and Carcinogenesis Studies of Sodium Fluoride in F344/N Rats and $B6C3F_1$ Mice (Drinking Water Studies). TR No. 393. National Toxicology Program, Washington, DC.

Pati, P.C., and S.P. Buhnya. 1987. Genotoxic effect of an environmental pollutant, sodium fluoride, in mammalian in vivo test systems. Caryologia 40:79-88.

Perry, W.G., F.A. Smith, and M.B. Kent. 1994. The halogens. Chapter 43 in Patty's Industrial Hygiene and Toxicology, Vol. II, Part F, G.F. Clayton and F.E. Clayton, eds. New York, NY: John Wiley & Sons, Inc.

Placke, M.E., M. Brooker, R. Persing, J. Taylor, and M. Hagerty. 1990. Final Report on Repeated-Exposure Inhalation Study of Hydrogen Fluoride in Rats. Battelle Memorial Institute, Columbus, OH. Docket No. OPPTS 42187, submitted by Battelle Washington Environmental Program, Arlington, VA.

Placke, M., and S. Griffin. 1991. Subchronic Inhalation Exposure Study of Hydrogen Fluoride in Rats. Batelle Memorial Institute, Columbus, OH. Docket No. OPPTS 42187, submitted by Battelle Washington Environmental Program, Arlington, VA.

Ronzani, E. 1909. Uber den einfluss der einatmungen von reizenden gasen der industrien auf die schutzkrafte des organismus gegenuber den infektiven krankheiten [in German]. Arch. Hyg. 70:217.

Rosenholtz, M.J., T.R. Carson, M.H. Weeks, F. Wilinski, D.F. Ford, and F.W. Oberst. 1963. A toxicopathologic study in animals after brief single exposures to hydrogen fluoride. Amer. Ind. Hyg. Assoc. J. 24:253-261.

Sadilova, M.S., K.P. Selyankina, and O.K. Shturkina. 1965. Experimental studies on the effect of hydrogen fluoride on the central nervous system. Hyg. Sanit. 30:155-160.

Stavert, D.M., D.C. Archuleta, M.J. Behr, and B.E. Lehnert. 1991. Relative acute toxicities of hydrogen fluoride, hydrogen chloride, and hydrogen bromide in nose- and pseudo-mouth-breathing rats. Fundam. Appl. Toxicol. 16:636-655.

Stokinger, H.E. 1949. Toxicity following inhalation of fluorine and hydrogen fluoride. Chapter 17 in Pharmacology and Toxicology of Uranium Compounds, C. Voegtlin and H.C. Hodge, eds. New York: McGraw-Hill Book Company.

ten Berge, W.F., A. Zwart, and L.M. Appleman. 1986. Concentration-time mortality response relationship of irritant and systemically acting vapors and gases. J. Hazard. Mater. 13:301-310.

Trevino, M.A. 1991. Hydrofluoric acid exposures: Long-term effects [draft]. Draft report cited in EPA 1993.

TNO/RIVM (The Netherlands Organization for Applied Scientific Research/National Institute of Public Health and the Environment). 1996. Hydrogen Fluoride: Risk Assessment [draft]. Chemical Substances Bureau, The Hague, The Netherlands.

Tepperman, P.B. 1980. Fatality due to acute systemic fluoride poisoning following a hydrofluoric-acid skin burn. J. Occup. Med. 22:691-692.

Vernot, E.H., J.D. MacEwen, C.C. Haun, and E.R. Kinkead. 1977. Acute toxicity and skin corrosion data for some organic and inorganic compounds and aqueous solutions. Toxicol. Appl. Pharmacol. 42:417-423.

Voroshilin, S.I., E.G. Plotko, and V. Ya Nikiforova. 1975. Mutagenic effect of hydrogen fluoride on animals. Tsitol. Genet. 9:40-42.

Waldbott, G.L., and J.R. Lee. 1978. Toxicity from repeated low-grade exposure to hydrogen fluoride - case report. Clin. Toxicol. 13:391-402.

Wing, J.S., L.M. Sanderson, J.D. Brender, D.M. Perrotta, and R.A. Beauchamp. 1991. Acute health effects in a community after a release of hydrofluoric acid. Arch. Environ. Health 46:155-160.

Wohlslagel, J., L.C. DiPasquale, and E.H. Vernot. 1976. Toxicity of solid rocket motor exhaust: Effects of HCl, HF, and alumina on rodents. J. Combust. Toxicol. 3:61-69.

Weiss, G. 1980. Hazardous Chemicals Data Book. Park Ridge, NJ: Noyes Data Corporation.

APPENDIX A

Time-Scaling Calculations for Hydrogen Fluoride AEGLs

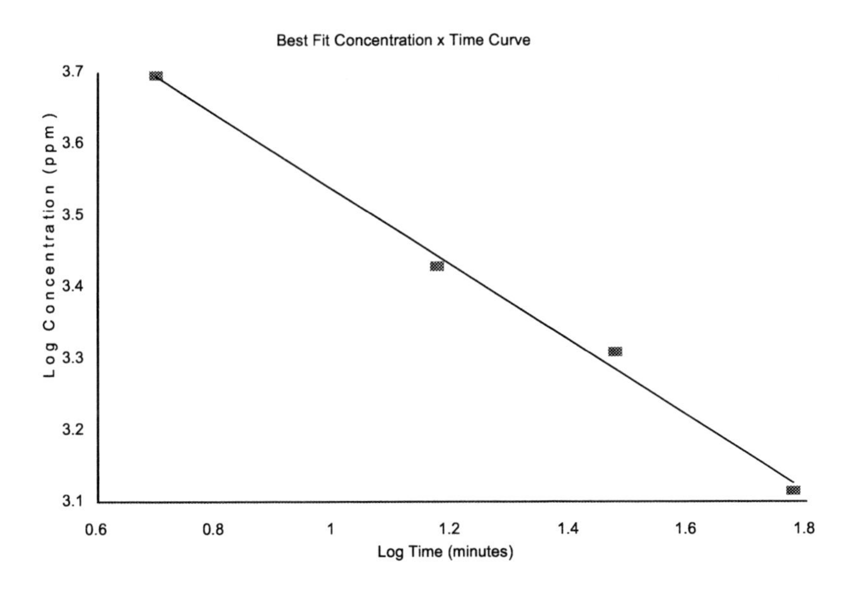

Data (LC_{50} values) (Rosenholtz et al. 1963)

Time (min)	Concentration (ppm)	Log time	Log concentration
5	4,970	0.6990	3.6964
15	2,689	1.1761	3.4296
30	2,042	1.4771	3.3101
60	1,307	1.7782	3.1163

Regression Output:

Intercept	4.0627
Slope	-0.5260
R Squared	0.9948
Correlation	-0.9974
Degrees of Freedom	2
Observations	4

$n = 1.9$
$k = 5.3E+07$

APPENDIX B

Derivation of AEGL Values

Derivation of AEGL-1

Key study:	Lund et al. 1997, 1999
Toxicity end point:	Biomarkers of exposure during 1 h exposure of exercising human subjects to several ranges of concentrations.
Time-scaling:	Not applied; adaptation occurs to the slight effects characterized by the AEGL-1.
Uncertainty factors:	3 for differences in human susceptibility. The resulting concentration should be protective of asthmatic individuals because it is below the average (2 ppm) and ranges of concentrations (up to 8.1 ppm) (Largent 1960, 1961) that produced slight to mild irritation in healthy adult male subjects.
Calculations:	The 3 ppm concentration was divided by the intraspecies UF of 3. The resulting concentration, 1 ppm, was used for all AEGL-1 time points.

Derivation of AEGL-2

Key studies:	Dalbey 1996; Dalbey et al. 1998a; Rosenholtz et al. 1963
Toxicity end point:	Lower respiratory tract effects (10-min value)—the 10-min NOAEL of 950 ppm in orally cannulated rats (Dalbey 1996) Irritation (30-min and 1-, 4-, and 8-h values)—signs of blinking, sneezing, and coughing in dogs exposed at 243 ppm for 1 h (Rosenholtz et al. 1963)

Time-scaling: $C^2 \times t = k$, based on the data of Rosenholtz et al. (1963), where $C = 243$ ppm, $t = 60$ min, UF = 10, and $n = 2$

$C^2/10 \times t = k$
$k = 35429.4$ ppm^2·min

Uncertainty factors:

10-min AEGL-2
Combined uncertainty factor of 10
3 for interspecies (effects are unlikely to differ between species; LC_{50} values were similar among species; oral cannulation maximizes the dose to the lower respiratory tract)
3 for intraspecies (oral cannulation maximizes the dose to the lower respiratory tract and is a potentially realistic model for human response to corrosive gases)

30-min and 1-, 4-, and 8-h AEGL-2
Combined uncertainty factor of 10
3 for interspecies (the dog is a sensitive species for sensory irritation)
3 for intraspecies

10-min AEGL-2: 950 ppm/10 = 95 ppm

30-min AEGL-2: $C^2/10 \times 30$ min = 35429.4 ppm^2·min
$C = 34.4$ ppm

1-h AEGL-2: $C = 243$ ppm/10 = 24.3 ppm

4-h AEGL-2: $C^2/10 \times 240$ min = 35429.4 ppm^2·min
$C = 12.2$ ppm

8-h AEGL-2: 12 ppm

Time-scaling to the 8-h exposure duration results in a value inconsistent with the human data. Because humans suffered no greater effect than slight irritation during intermittent exposures at 8.1 ppm on a repeated basis (Largent 1961, 1962), the calculated concentration of 8.6 ppm was considered too low. Therefore, the 8-h AEGL-2 value was set equal to the 4-h value.

Derivation of AEGL-3

Key studies: Dalbey 1996; Dalbey et al. 1998a; Wohlslagel et al. 1976

Toxicity end point: Lethality (10-min value)—LC_{05} in rats (1,764 ppm) (Dalbey 1996)
Lethality (30-min and 1-, 4-, and 8-h values)—1-h no-death value in the mouse (263 ppm) (Wohlslagel et al. 1976)

Time-scaling: $C^2 \times t = k$, based on the data of Rosenholtz et al. (1963), where $C = 263$ ppm, $t = 60$ min, UF/MF = 6, and $n = 2$

$(263 \text{ ppm}/6)^2 \times 60 \text{ min} = 115{,}281.67 \text{ ppm}^2\cdot\text{min}$

Uncertainty factors: *10-min AEGL-3*
Combined uncertainty factor of 10
3 for interspecies (effects are unlikely to differ greatly among species; LC_{50} values were similar among species; oral cannulation maximizes the dose to the lower respiratory tract)
3 for intraspecies (oral cannulation maximizes the dose to the lower respiratory tract and is a potentially realistic model for human response to corrosive gases)

30-min and 1-, 4-, and 8-h AEGL-3
1 for interspecies (the mouse was the most sensitive species; LC_{50} values differed by approximately 3 between the rat and mouse and effects are unlikely to differ between species).
3 for intraspecies

Modifying factor: 2 to account for the fact that the highest nonlethal value was close to the LC_{50} (applied to the 30-min, and 1-, 4-, and 8-h values)

10-min AEGL-3: 1,700 ppm (rounded from 1,764)/10 = 170 ppm

30-min AEGL-3: $C^2/6 \times 30$ min = 115,281.67 ppm^2·min
C = 61.9 ppm

1-h AEGL-3: C = 263 ppm/6 = 43.8 ppm

4-h AEGL-3: $C^2/6 \times 240$ min = 115,281.67 ppm^2·min
C = 21.9 ppm

8-h AEGL-3: $C^2/6 \times 480$ min = 115,281.67 ppm^2·min
C = 15.4 ppm (set equal to the 4-h value of 22 ppm)

The time-scaled 8-h AEGL-2 value of 15 ppm is considered inconsistent with the animal data. Rats survived for 8 d during a 14-d exposure at 5 ppm (Placke et al. 1990). No deaths occurred in groups of four male and female rhesus monkeys inhaling 690 ppm for 1 h (MacEwen and Vernot 1970). In a longer-term study, two rhesus monkeys survived a 50-d exposure at 18.5 ppm, 6-7 h/d for 50 d for a total of 309 exposure hours (Machle and Kitzmiller 1935). Therefore, the 8-h AEGL-3 was set equal to the 4-h AEGL-3.

APPENDIX C

ACUTE EXPOSURE GUIDELINE LEVELS FOR HYDROGEN FLUORIDE (CAS No. 7664-39-3)

DERIVATION SUMMARY

AEGL-1				
10 min	30 min	1 h	4 h	8 h
95 ppm	34 ppm	24 ppm	12 ppm	12 ppm
Key references: (1)Lund et al. 1997. Exposure to hydrogen fluoride: an experimental study in humans of concentrations of fluoride in plasma, symptoms, and lung function. Occup. Environ. Med. 54:32-37. (2) Lund et al., 1999. Increased CD3 positive cells in bronchoalveolar lavage fluid after hydrogen fluoride inhalation. Scand. J. Work. Environ. Health 25:326-334.				
Test species/strain/number: 20 healthy male volunteers				
Exposure route/concentrations/durations: Inhalation; average concentrations of 0.2-0.7 ppm (n = 9), 0.85-2.9 ppm (n = 7), and 3.0-9.3 ppm (n = 7).				
Effects: At 0.2-0.7 ppm, no to low sensory irritation; no change in FVC, FEV_1; no inflammatory response in bronchoalveolar lavage fluid (BAL). At 0.85-2.9 ppm, no to low sensory irritation; no change in FVC, FEV_1; increase in the percentage of CD3 cells and myeloperoxidase in bronchial portion of BAL; no increases in neutrophils, eosinophils, protein, or methyl histamine in BAL. At 3.0-6.3 ppm, low sensory irritation; no change in FVC, FEV_1; increase in the percentage of CD3 cells and myeloperoxidase in bronchial portion of BAL; no increases in neutrophils, eosinophils, protein, or methyl histamine in BAL.				
End point/concentration/rationale: Subthreshold concentration for inflammation of 3 ppm (0.85-2.9 ppm) for 1 h, which was without sensory irritation was chosen as the basis for the AEGL-1.				
Uncertainty factors/rationale: Total uncertainty factor: 3 Interspecies: Not applicable since human subjects were the test species. Intraspecies: 3. The subjects were healthy adult males. The resulting concentration is far below tested concentrations that did not cause symp toms of bronchial constriction in healthy adults (ranges up to 6.3 ppm [Lund et al. 1997] and 8.1 ppm [Largent 1960, 1961])				

AEGL-1 *Continued*
Modifying factor: Not applicable
Animal to human dosimetric adjustment: Not applicable, human data used.
Time-scaling: Not applied. AEGL-1 values were calculated by adjusting the 1-h concentration of 3 ppm by a UF of 3. Because the response to slight irritation would be similar at shorter exposure durations, the 10- and 30-min values were set equal to the 1-h concentration.
Data adequacy: The values are supported by the earlier study of Largent (1969, 1961) in which five healthy human volunteers were exposed at 1.42-4.74 ppm for 10 to 50 d with no greater effects than slight irritation and reddened facial skin. Effects were no more severe in two individuals who were exposed at concentrations up to 7.9 ppm and 8.1 ppm during some of the exposure days.

AEGL-2				
10 min	30 min	1 h	4 h	8 h
95 ppm	34 ppm	24 ppm	12 ppm	12 ppm
Key references: (1) Dalbey, W. 1996. Evaluation of the toxicity of hydrogen fluoride at short exposure times. Petroleum Environmental Research Forum Project 92-09. Performed at Stonybrook Laboratories, Inc., Pennington, NJ. (2) Dalbey, W., B. Dunn, R. Bannister, W. Daughtrey, C. Kirwin, F. Reitman, A. Steiner, and J. Bruce. 1998. Acute effects of 10-minute exposure to hydrogen fluoride in rats and derivation of a short-term exposure limit for humans. Regulat. Toxicol. Pharmacol. 27:207-216. (3) Rosenholtz, M.J., T.R. Carson, M.H. Weeks, F. Wilinski, D.F. Ford and F.W. Oberst. 1963. A toxicopathologic study in animals after brief single exposures to hydrogen fluoride. Amer. Ind. Hyg. Assoc. J. 24:253-261.				
Test species/strain/gender/number: female Sprague-Dawley rats, 20/exposure group (Dalbey 1996); mongrel dogs, 2/exposure group (Rosenholtz et al. 1963)				
Exposure route/concentrations/duration: 10-min inhalation exposures of orally cannulated rats to 135, 271, 950, or 1,764 ppm (Dalbey, 1996); 60-min inhalation exposures of mongrel dogs to 157 ppm or 243 ppm (Rosenholtz et al. 1963)				
Effects: In the 10-min inhalation exposures of orally cannulated rats (Dalbey 1996), at 135 ppm there was no effect; at 271 ppm there was no effect; 950 ppm was set as a no-observed-adverse-effect level (NOAEL) (increase in myeloperoxidase and polymorphonuclear leukocytes in BAL); and 1,764 ppm resulted in death of one of 20 animals. In the 60-min inhalation exposures of mongrel dogs (Rosenholtz et al. 1963), at 157 ppm there was mild eye irritation and sneezing and dry cough that persisted for 2 d; and at 243 ppm there was eye, nasal, and respiratory irritation (blinking, sneezing, and coughing during exposures; cough persisted for 2 d and during exercise for up to 10 d).				
End point/concentration/rationale: For the 10-min exposure, the NOAEL of 950 ppm in orally cannulated rats was chosen because it addresses the relevant exposure period and represents the highest concentration tested that did not result in death. In addition, direct delivery of hydrogen fluoride to the trachea via cannulation is a sensitive model and simulates 100% mouth-breathing in humans exposed to irritant gases. For longer-term exposures, the study by Rosenholtz et al. (1963) was chosen because it was well designed and used dogs which represent a sensitive model for irritants. The highest exposure				

AEGL-2 *Continued*
level of 243 ppm for 60 min, which resulted in symptoms/effects of great discomfort but is not expected to impair the ability to escape or result in irreversible or long-lasting effects, was chosen as the threshold for AEGL-2 effects.
Uncertainty factors/rationale: *10-min AEGL-2 values* Total uncertainty factor: 10 Interspecies: 3. A sensitive model was used (orally cannulated rats). Intraspecies: 3. Oral cannulation maximizes the dose to the lungs and is relevant to mouth-breathing humans. *30-min and 1-, 4-, and 8-h AEGL-2 values* Total uncertainty factor: 10 Interspecies: 3. A sensitive species was used (other studies with irritant gases show an irritant response in the dog at concentrations that are nonir ritating to rodents). Intraspecies: 3. A greater factor would lower the value to concentrations that were non-irritating in human studies.
Modifying factor: Not applicable
Animal to human dosimetric adjustment: Insufficient data
Time-scaling: $C^n \times t = k$ where $n = 2$ was derived based on regression analysis of rat LC_{50} studies conducted at time periods of 5, 15, 30, and 60 min. A second study using rabbits and guinea pigs and conducted over time periods of 5 min to 6 h resulted in the same value for n (reported in a third study). End points for the second study were both irritation and death. Because the time-scaled 8-h value of 8.6 ppm was inconsistent with the data of Largent (1960, 1961), the 8-h AEGL-2 was set equal to the 4-h value.
Data adequacy: Based on the following observations, there is considerable support for the scientific credibility of the AEGL-2 values. Oral cannulation bypasses nasal scrubbing and maximizes the dose to the lung. Two species (rat and dog) were tested but not at the same concentrations. A similar irritant response was observed in the rat at higher test concentrations. According to Alarie (Environ. Health Persp. 42:9-13), one-tenth of the mouse RD_{50} for irritant chemicals can be tolerated for "hours" by humans. The mouse RD_{50} is 151 ppm; deriving AEGL-2 values based on the RD_{50} results in a human 4- or 8-h exposure of 15 ppm, slightly higher than the AEGL-2 values based on irritant effects in the dog. The database for irritant effects of hydrogen fluoride is extensive. Five species were tested over a range of concentrations for time periods of 5 min to 6 h.

AEGL-3				
10 min	30 min	1 h	4 h	8 h
170 ppm	62 ppm	44 ppm	22 ppm	22 ppm

Key References: (1) Dalbey, W. 1996. Evaluation of the toxicity of hydrogen fluoride at short exposure times. Petroleum Environmental Research Forum Project 92-09. Performed at Stonybrook Laboratories, Inc., Pennington, NJ.
(2) Dalbey, W., B. Dunn, R. Bannister, W. Daughtrey, C. Kirwin, F. Reitman, A. Steiner, and J. Bruce. 1998. Acute effects of 10-minute exposure to hydrogen fluoride in rats and derivation of a short-term exposure limit for humans. Regulat. Toxicol. Pharmacol. 27:207-216.
(3) Wohlslagel, J., L.C. DiPasquale and E.H. Vernot. 1976. Toxicity of solid rocket motor exhaust: Effects of HCl, HF, and alumina on rodents. J. Combust. Toxicol. 3:61-69.

Test species/strain/gender/number: female Sprague-Dawley rats, 20/exposure group; female CF-1 mice, 10/exposure group

Exposure route/concentrations/durations: 10-min inhalation exposures of orally cannulated rats at 135, 271, 950, or 1,764 ppm. 60-min inhalation exposures of female mice at 263, 278, 324, 381, or 458 ppm

Effects: In the 10-min inhalation exposures of orally cannulated rats, at 135 ppm there were no effects; at 271 ppm there was no effect; 950 ppm was set as a no-observed-adverse-effect level (NOAEL) (increase in myeloperoxidase and polymorphonuclear leukocytes in BAL); and 1,764 ppm resulted in the death of one of 20 animals. In the 60-min inhalation exposures of female mice, at 263 ppm there were no deaths; at 278 ppm there were 1/10 deaths; at 324 ppm there were 7/10 deaths; at 381 ppm there were 6/10 deaths; at 458 ppm there were 9/10 deaths.

End point/concentration/rationale: For the 10-min AEGL-3, the LC_{05} of 1,764 ppm was rounded down to 1,700 ppm. Although 1/20 deaths is higher than the usual threshold for the AEGL-3 (1/100 deaths), the oral cannulation model is conservative compared with normal nose breathing as it bypasses nasal scrubbing and maximizes the dose to the lung. No higher concentrations were tested at the 10-min exposure period. The concentration resulting in no deaths in the mouse, 263 ppm, was chosen for the longer exposure periods. This specific data set was selected because, based on LC_{50} values in several studies, the mouse was the most sensitive of three tested species (monkey, rat, and mouse).

Uncertainty factors/rationale:
10-min AEGL-3 values
Total uncertainty factor: 10

AEGL-3 *Continued*
Interspecies: 3. Based on LC_{50} values in the same studies, the rat was approximately three times less sensitive than the mouse to the lethal effects of hydrogen fluoride; however, the delivery of hydrogen fluoride directly to the trachea via oral cannulation is a conservative model. Intraspecies: 3. Oral cannulation maximizes the dose to the lungs and is relevant to mouth breathing humans. *30-min and 1-, 4-, and 8-h AEGL-2 values* Total uncertainty factor: 3 Interspecies: 1. Based on LC_{50} values, the mouse was the most sensitive of three tested species; of several studies involving the mouse, this study had the lowest lethal values. Intraspecies: 3. Application of a greater uncertainty factor would reduce concentrations to those found only slightly irritating in human studies.
Modifying factor: For 30-min and 1-, 4-, and 8-h AEGL-2 values, 2. The highest non-lethal value was close to the LC_{50} value
Animal to human dosimetric adjustment: Insufficient data
Time-scaling: $C^n \times t = k$ where $n = 2$ based on regression analysis of rat LC_{50} studies conducted at time periods of 5, 15, 30, and 60 min. A second study using rabbits and guinea pigs and conducted over time periods of 5 min to 6 h resulted in the same value for n (reported in a third study). End points for the second study were both irritation and death. Because the time-scaled 8-h AEGL-3 value of 15 ppm was inconsistent with results of longer-term studies with monkeys and rodents, the 8-h value was set equal to the 4-h value.
Data adequacy: There is considerable support for the AEGL-3 values as the database for hydrogen fluoride is extensive with multiple studies of lethality conducted at several exposure durations and involving five species of mammals (monkey, rat, mouse, guinea pig, and rabbit). Studies with multiple dosing regimens generally showed a clear dose-response relationship. A few longer-term studies were also available and served as supporting data. Tissue and organ pathology indicated that the toxic mechanism was the same across species. Difficulties in maintaining/measuring exposure concentrations were encountered in some of the studies; studies in which these difficulties were described were not used to derive AEGL values.

4

Toluene 2,4- and 2,6-Diisocyanate[1]

Acute Exposure Guideline Levels

SUMMARY

Toluene diisocyanate (TDI) is among a group of chemicals, the isocyanates, that are highly reactive compounds containing an -NCO group. TDI exists as both the 2,4- and 2,6- isomers, which are available commercially, usually in ratios of 65:35 or 80:20 (Karol 1986; WHO 1987). TDI has been used in the manufacture of polyurethane foam products as well as paints, varnishes, elastomers, and coatings (WHO 1987).

Inhaled TDI causes irritation and sensitization of the respiratory tract. Sensitization may occur from either repeated exposure over a relatively long

---

[1]This document was prepared by the AEGL Development Team comprising Carol Wood (Oak Ridge National Laboratory) and National Advisory Committee (NAC) on Acute Exposure Guideline Levels for Hazardous Substances member Steven Barbee (Chemical Manager). The NAC reviewed and revised the document and AEGL values as deemed necessary. Both the document and the AEGL values were then reviewed by the National Research Council (NRC) Subcommittee on Acute Exposure Guideline Levels. The NRC subcommittee concludes that the AEGLs developed in this document are scientifically valid conclusions on the basis of the data reviewed by the NRC and are consistent with the NRC guidelines reports (NRC 1993, 2001).

period of time (i.e., years), or it may consist of an induction phase precipitated by a relatively high concentration followed by a challenge phase in which sensitized individuals react to extremely low concentrations of TDI. Only irritation effects were considered in establishing AEGL values, because sensitized individuals are considered to be hypersusceptible. Although individuals with existing TDI sensitization are present in the general population, that presensitization cannot be estimated. If the number of individuals sensitized to TDI in the general population were quantifiable, a different approach to derivation of AEGL values might have been considered. At any of the AEGL levels, there might be individuals who have a strong reaction to TDI, and those individuals might not be protected within the definition of effects for each level.

Human data were available for the derivation of AEGL-1 and AEGL-2. Fifteen asthmatic subjects were exposed to TDI at 0.01 parts per million (ppm) for 1 hour (h), and then after a rest of 45 minutes (min), they were exposed at 0.02 ppm for 1 h. A nonasthmatic referent group of 10 individuals was exposed at 0.02 ppm for 2 h (Baur 1985). None of the individuals had a history of isocyanate exposure, and the asthmatic subjects were not sensitized to TDI. Although no statistically significant differences in lung function parameters were observed among asthmatic subjects during or after exposure, nonpathological bronchial obstruction was indicated in several individuals. In the referent group, there was a significant increase in airway resistance immediately and at 30 min after the initiation of exposure, but none of the subjects developed bronchial obstruction. Both groups reported eye and throat irritation, cough, chest tightness, rhinitis, dyspnea, and/or headache, but time to onset of symptoms was not given. There was also no indication whether symptoms were more severe in asthmatic subjects that inhaled 0.01 or 0.02 ppm. Therefore, the 0.02-ppm concentration was identified as the basis for the 10-min, 30-min, and 1-h AEGL-1 values. The 0.01-ppm concentration was identified as the basis for the 4- and 8-h AEGL-1 values. It should be noted that the AEGL-1 values are below a reported odor detection threshold of 0.05 ppm (Henschler et al. 1962).

Derivation of AEGL-2 was based on human data. Exposure of volunteers to TDI at 0.5 ppm for 30 min resulted in severe eye and throat irritation and lacrimation (Henschler et al. 1962). A higher exposure concentration was intolerable. Extrapolations were made using the equation $C^n \times t = k$ (C = concentration, t = time, and k is a constant), where n ranges from 0.8 to 3.5 (ten Berge et al. 1986). In the absence of an empirically derived, chemical-specific exponent, scaling was performed using $n = 3$ for extrapolating to the 10-min time point and $n = 1$ for the 1-h and 4-h time points.

The 4-h value was used for the 8-h value, because extrapolation to 8 h resulted in a concentration similar to that shown to be tolerated for >7 h with only mild effects. An uncertainty factor (UF) of 3 was applied to account for sensitive individuals; use of a greater UF results in values below those supported by human data for AEGL-2 effects.

No human data were available for derivation of AEGL-3 values. Human fatalities attributed to TDI-induced chemical pneumonitis have occurred under unusual circumstances. Exposure concentrations in those accidents were not measured. Therefore, animal data were used to derive AEGL-3 values. On the basis of LC_{50} values (concentrations lethal to 50% of subjects), the species most sensitive to the effects of TDI is the mouse. The 4-h mouse LC_{50} of 9.7 ppm (Duncan et al. 1962) was divided by 3 to estimate a threshold of lethality based on the regression plot of mortality vs concentration. This estimated 4-h lethality threshold was used to extrapolate to the 30-min and 1- and 8-h AEGL-3 time points. Values were scaled using the equation $C^n \times t = k$, where n ranges from 0.8 to 3.5 (ten Berge et al. 1986). In the absence of an empirically derived, chemical-specific exponent, scaling was performed using $n = 3$ for extrapolating to the 30-min and 1-h time points and $n = 1$ for the 8-h time point. A total UF of 10 was applied, which includes 3 to account for sensitive individuals and 3 for interspecies extrapolation (use of a greater UF would result in values below those supported by human data for AEGL-3 effects). According to Section 2.7 of the standing operating procedures for the derivation of AEGLs (NRC 2001), 10-min values are not to be scaled from an experimental exposure time of ≥4 h. Therefore, the 30-min AEGL-3 value was also adopted as the 10-min value.

1. INTRODUCTION

TDI is among a group of chemicals, the isocyanates, that are highly reactive compounds containing an -NCO group. TDI exists as both 2,4- and 2,6- isomers, which are available commercially, usually in ratios of 65:35 or 80:20 (Karol 1986; WHO 1987). An estimated 1,225 million pounds of TDI were produced in 2000, and greater than 90% was used in the manufacture of flexible urethane foams (CPS 2001). TDI is produced from the reaction of diaminotoluenes with phosgene in a closed system, and TDI has also been used in the manufacture of urethane paints, varnishes, elastomers, and coatings. The chemical may dimerize slowly at ambient temper-

TABLE 4-1 Summary of AEGLs Values for Toluene 2,4- and 2,6-Diisocyanate (ppm [mg/m^3])

Classification	10 min	30 min	1 h	4 h	8 h	End Point (Reference)
AEGL-1 (Nondisabling)	0.02 (0.14)	0.02 (0.14)	0.02 (0.14)	0.01 (0.07)	0.01 (0.07)	Chest tightness, eye and throat irritation (Baur 1985)
AEGL-2 (Disabling)	0.24 (1.71)	0.17 (1.21)	0.083 (0.59)	0.021 (0.15)	0.021 (0.15)	Severe eye and throat irritation, lacrimation (Henschler et al. 1962)
AEGL-3 (Lethal)	0.65 (4.63)	0.65 (4.63)	0.51 (3.63)	0.32 (2.28)	0.16 (0.93)	4-h LC_{50} in the mouse (Duncan et al. 1962)

Abbreviations: mg/m^3, milligrams per cubic meter; ppm, parts per million.

atures and more rapidly at higher temperatures, and trimerization occurs at 100-200°C (WHO 1987).

The odor threshold for 2,4- and 2,6-TDI was found to be 0.05 ppm (Henschler et al. 1962). In early human primary irritation testing with 2,4-TDI, 50% of subjects reported the least detectable odor at 0.4 ppm. Irritation of the nose and throat occurred at 0.5 ppm, and an appreciable odor was noted at 0.8 ppm (Zapp 1957; Wilson and Wilson 1959).

Toxicological effects to the respiratory tract from inhaled TDI may be divided into two distinct categories: (1) primary irritation and (2) immunologic hypersensitivity. The chemically reactive isocyanate group has been suggested as the cause of both effects. The primary irritation associated with inhaled TDI is a nonspecific inflammatory response characteristic of that produced by other primary irritants. Inflammation is also a consequence of sensitization, but it is caused by an immunologically mediated reaction leading to antibody formation (Karol 1986; WHO 1987), and that response is individual-specific. Sensitization consists of an induction phase precipitated by a relatively high concentration, followed by a challenge phase in which immunologically sensitized individuals react to extremely low concentrations of TDI that are, in some persons, below the current ACGIH Threshold Limit Value (TLV) of 5 ppb. Some studies showed detection of IgE antibodies in sensitized individuals, although others found variable or negative results (Karol 1986). IgG antibodies specific to TDI have been detected in both asymptomatic and symptomatic workers (Baur 1985). The immune-mediated inflammatory response of the respiratory

TABLE 4-2 Physicochemical Data for Toluene Diisocyanate

Parameter	Value	Reference
Synonyms	TDI; tolylene diisocyanate	Budavari et al. 1996
CAS registry no.	584-84-9 (2,4-TDI) 91-08-7 (2,6-TDI)	
Chemical formula	$C_9H_6N_2O_2$	Budavari et al. 1996
Molecular weight	174.16	Budavari et al. 1996
Physical state	Clear yellow liquid	Shiotsuka 1987b
Vapor pressure	0.011 mm Hg at 25°C 3.2 mm Hg at 100°C	Woolrich 1982
Vapor density (air = 1)	6.0	ACGIH 1991
Specific gravity	1.22 g/cm^3	Shiotsuka 1987b
Boiling/flash point	251°C/132°C (open cup)	ACGIH 1991
Solubility in water	Reacts with water	ACGIH 1991
Conversion factors	1 ppm = 7.12 mg/m^3 1 mg/m^3 = 0.14 ppm	Hartung 1994

tract has been characterized by persistent activation of lymphocytes, chronic expression of certain cytokines (Maestrelli et al. 1995), neutrophilia, eosinophilia (Fabbri et al. 1987), and decreased lymphocyte cAMP levels (Butcher et al. 1979).

The physicochemical properties of TDI are given in Table 4-2 (above).

2. HUMAN TOXICITY DATA

2.1. Acute Lethality

Human fatalities from TDI exposure are not common. Accidents have involved unusual circumstances. TDI concentrations were not measured, and the isomer was not always identified. In one case, a worker was trapped in a room following the explosion of a storage vessel. The victim was unconscious for an unknown but extended exposure duration (Horspool and Doe 1977). A second report involved a salvage diver who was blowing polyurethane foam in the hold of a ship. When his air supply failed, he removed his mask and was exposed to an atmosphere containing a high (but unknown) concentration of Freon and 2,4-TDI. The diver was unconscious and submerged in sea water when rescued. He died 4 days (d) later despite extensive resuscitative efforts (Linaweaver 1972). Deaths in both of those cases appeared to be due to pulmonary edema subsequent to chemical pneumonitis.

2.2. Nonlethal Toxicity

2.2.1. Case Reports

A 50-y-old male was drenched in TDI (isomer mixture not specified) when a hose detached from a tanker truck he was helping to unload. The individual had no history of respiratory illness, asthma, or allergic disease. Shortly after exposure, he developed shortness of breath, wheezing, and cough. Evaluation 12 y later showed persistent asthma and variable airway obstruction despite no further exposure to isocyanates. However, his asthma became more severe after exposure to other irritants in the workplace. An asthmatic attack was provoked by challenge with 10 ppb (71.2 $\mu g/m^3$) TDI for 8 min (Moller et al. 1986).

In contrast to the above report, TDI sensitivity was lost in a worker 11 months (mo) after removal from exposure, and nonspecific bronchial hyperresponsiveness resolved after 17 mo despite the continued presence of serum IgE antibodies (Butcher et al. 1982). The TDI isomer was not reported for either the occupational exposure or the experimental challenge testing.

Isocyanate vapor concentrations have been measured to estimate worker exposure during the spray application of polyurethane foam. Personal samplers were attached to the sprayers, but the exact location of the samplers (i.e., breathing zone) was not specified. Average exposure concentrations ranged from 0.021 ppm to 0.045 ppm, and exposure durations ranged from 105 min to 442 min. No head or eye protection was provided except for the voluntary use of plastic bags over the heads of the sprayers. Reddening of the eyes and lacrimation were observed in "numerous" workers during the course of the study (Hosein and Farkas 1981). This study neither identified the isocyanate isomers nor correlated the prevalence of clinical signs in the workers and the exposure concentrations and durations.

Case reports of TDI intoxication at 15 plants involved in polyurethane operations (most likely involving mixed isomers) were investigated by the Massachusetts Department of Labor and Industries (Elkins et al. 1962).

Workers complained of eye and throat irritation, tightness of the chest, nausea and vomiting, nonproductive cough, and restlessness despite the control of TDI vapor concentrations according to the standard in effect at that time. Milder effects were documented in plants with maximum workroom concentrations at 0.02 ppm, and more severe effects were documented in plants with maximum workroom concentrations at ≥0.07 ppm. The authors concluded that the maximum allowable concentration of 0.1 ppm was too high and recommended that a limit of 0.01 ppm be adopted. Current occupational exposure standards are given in Section 8.2 of this document.

Workers in a manufacturing plant involved in the production of isocyanate foam complained of coughing, sore throat, dyspnea, fatigue, and night sweats (Hama 1957). A change in the manufacturing process placed workers in a poorly ventilated room, which resulted in symptoms in 12 of 12 workers. Isocyanate concentrations (isomer not specified) ranging from 0.03 ppm to 0.07 ppm were measured in the room (assumed to be area samples). Following the return to previous manufacturing processes, no complaints or symptoms of exposure have occurred, and measured concentrations of isocyanates were found to be <0.03 ppm.

Seven men developed cough, dyspnea, chest pain, wheezing, and hemoptysis following exposure to a plastic varnish containing TDI (isomer not specified). Air samples taken in the work area—after temporary measures had been implemented to improve ventilation—contained 0.08 ppm to 0.1 ppm. Six of the seven individuals had varying degrees of respiratory impairment, as determined by timed vital capacity. Improvement was noted in five when reexamined at 2-2.5 mo after exposure (Maxon 1964).

2.2.2. Epidemiologic Studies

Numerous occupational studies have evaluated pulmonary function in workers exposed to TDI. However, most have failed to account for confounding factors such as smoking status, sampling methods that failed to detect and quantify both isomers, high rates of annual FEV_1 (forced expiratory volume in 1 second) decline in control populations, and high intra- and interindividual variation in lung-function testing (EPA 1996). A study by Diem et al. (1982, as cited in EPA 1996) accounted for those factors and followed TDI production workers prospectively over a 5-y period. Investigators identified two exposure groups, defined as low and high, with arithmetic mean concentrations for never-smokers of 0.9 ppb and 1.9 ppb (6.41 $\mu g/m^3$ and 13.53 $\mu g/m^3$), respectively. Never-smokers in the high TDI exposure category had a significant ($p \leq 0.001$) decline in FEV_1 and forced expiratory flow at 25-75% when compared with never-smokers in the low-exposure category. Similar results in FEV_1 were found when the same groups were recategorized on the basis of time spent inhaling workplace air containing a concentration above 20 ppb (142.4 μg/m^3). The U.S. Environmental Protection Agency (EPA) (1996) used the Diem et al. (1982, as cited in EPA 1996) study to calculate a reference concentration (RfC) of 0.98×10^{-5} ppm. EPA (1996) concluded that "[a]lthough the mean exposure values determined in this study are close to the detection limit of the sampling and detection method, the values are considered accurate because they were obtained by continuous monitoring over the entire workday."

One of the largest occupational studies of polyurethane foam workers was conducted by Bugler et al. (1991). That 5-y study was designed to investigate the risk of sensitization to isocyanates and the longitudinal change in ventilatory capacity. Personal exposures were measured using modified MCM paper tape monitors. Low, intermediate, and high exposure groups were identified with average TDI exposures of 0.3, 0.6, and 1.2 ppb (2.14, 4.27, and 8.54 μg/m^3), respectively. There were no significant effects of exposure as measured by changes in the rate of decline in several parameters of pulmonary function. Over 5 y, the rate of sensitization among the original subjects was 3.1%, or 0.6% per year. Of note is the 4% rate of sensitization among new hires. Overall, in 47% of workers diagnosed, sensitization occurred after exposure to TDI concentrations less than 20 ppb (142.4 μg/m^3) (Bugler et al. 1991). A major problem with this study was that the limit of detection was only 4 ppb (28.48 μg/m^3), indicating that estimates of cumulative daily exposures were based on measurements below the limit of quantitation (Garabrant and Levine 1994).

A comprehensive review of the epidemiological studies on TDI was prepared by Garabrant and Levine (1994). Those authors concluded that respiratory sensitization occurs in less than 1% of subjects per year who are exposed to TDI at levels below 20 ppb (142 μg/m^3), and that sensitization is almost entirely attributable to short-term excursions above that level.

2.2.3. Experimental Studies

Provocative inhalation challenge tests using 2,4- and 2,6-TDI (80:20) were administered to 15 asthmatic subjects and 10 healthy controls (Baur 1985). None of the individuals had a history of isocyanate exposure, and the asthmatic subjects were not sensitized to TDI. All individuals classified as asthmatic had a history of asthmatic episodes and a significant response to acetylcholine challenge test. Asthmatic subjects were exposed to TDI at

0.01 ppm for 1 h, and then, after a rest of 45 min, they were exposed at 0.02 ppm (0.142 milligrams per cubic meter [mg/m^3]) for 1 h. Controls were exposed to TDI at 0.02 ppm for 2 h. In the control group, there was a statistically significant ($p \leq 0.05$) increase in airway resistance (R_{aw}) immediately after and 30 min after the beginning of exposure. For the asthmatic group, no statistically significant differences were observed from pretest group mean values for lung function parameters during or after exposure. However, eight of 15 individuals had an increase in R_{aw} of >50%, and four of those subjects had significant bronchial obstruction, which was defined as an increase in specific airway resistance of >50%. Specific airway resistance was calculated as the product of the R_{aw} multiplied by the intrathoracic gas volume. More important, among the asthmatics, no individual decrease in FEV_1 of more than 20% was observed. The increases in R_{aw} and decreases in FEV_1 are not considered pathologic for the asthmatic subjects because the changes were relatively minor and inconsistent within individuals. Individual values for R_{aw} and FEV_1 in several of the asthmatic subjects following TDI exposure are given in Table 4-3. Increases in R_{aw} did not correspond with decreases in FEV_1, and neither parameter could be used as an indication of reported discomfort. For example, individual 8 had the greatest increase in R_{aw} (3.2 times), but the FEV_1 showed essentially no decline (3.51 L vs 3.41 L). Individual 9 also showed no decline in FEV_1 (4.01 L vs 3.91 L) and had a 1.5-time increase in R_{aw}. Individual 5, who had the greatest decline in FEV_1, reported no symptoms of discomfort.

Five of the asthmatic individuals complained of chest tightness, rhinitis, cough, dyspnea, throat irritation, and/or headache during exposure; three controls reported eye irritation and/or cough. Some of the symptoms lasted for several hours post-exposure. The study author concluded that some people with pre-existing bronchial hyper-reactivity respond to TDI at or below the ACGIH short-term exposure limit (STEL = 0.02 ppm) (see Section 8.2) with bronchial obstruction (Baur 1985).

Henschler et al. (1962) exposed six healthy male volunteers to 2,4- and 2,6-TDI (65:35), 2,4-TDI, or 2,6-TDI at measured concentrations ranging from 0.01 ppm to 1.3 ppm for 30 min. The volunteers were exposed at all concentrations, but at only one concentration per day, and the concentrations were randomly selected. Volunteers had no prior knowledge of the isomer or concentration selected. The results are summarized in Table 4-4. The odor threshold was found to be 0.05 ppm. A concentration-dependent increase in sensory irritation was reported. There was slight eye and nose irritation at 0.1 ppm and marked discomfort at ≥0.5 ppm. 2,6-TDI appeared slightly more irritating than the 2,4- isomer, but was similar to the mixture.

TABLE 4-3 Increase in R_{aw} Compared with FEV_1 Following Exposure to Toluene Diisocyanate in Asthmatic Subjects

		FEV_1 (L)		
Individual	Maximum increase in R_{aw}	Before TDI Exposure	Lowest Value After TDI Exposure[a]	Symptoms
3	1.5×	3.0	2.5	Rhinitis, throat burning sensation, mild cough
5	2.0×	3.7	3.1	None
6	1.7×	4.8	4.4	Chest tightness
8	3.2×	3.5	3.4	Cough, chest tightness, dyspnea
9	1.5×	4.0	3.9	None

[a]None of the individuals experienced a >20% decline in FEV_1.
Source: Data from Baur 1985.

No adverse effects were reported in two healthy men exposed to 2,4- and 2,6-TDI (30:70) at up to 9.8 ppb (70 µg/m³) for 4 h (Brorson et al. 1991) or in five healthy men exposed to 2,4- and 2,6-TDI (65:35) at 5.6 ppb (40 µg/m³) for 7.5 h (Skarping et al. 1991). No further details of those studies were reported.

In 10 individuals with positive methacholine challenge tests, 2,4-TDI inhalation challenge testing at up to 20 ppb (142 µg/m³) for 15 min resulted in no change in FEV_1 (Moller et al. 1986). No further details of this study were reported.

Four adults with occupational asthma associated with exposure to isocyanates were challenged with TDI (isomer not specified), and their responses were assessed (Vandenplas et al. 1993). The duration of work exposure ranged from 7 to 17 y, and the duration of symptoms ranged from 0.5 to 10 y. Subjects were exposed at varying concentrations (5, 10, 15, and 20 ppb [35.6, 71.2, 106.8, 142.4 µg/m³]) for 1-90 min such that the $C \times t$ product remained constant. A positive asthmatic response was defined as a ≥20% drop in FEV_1. Although the effective $C \times t$ was highly variable between individuals (45-450 ppb·min), it remained constant for each person. Therefore, the authors concluded that both concentration and duration of exposure determined the occurrence of an asthmatic reaction in sensitized

TABLE 4-4 Effects of Controlled Inhalation Exposure to Toluene Diisocyanate in Volunteers[a]

Concentration (ppm)	Effect
0.01 or 0.02	2,4/2,6; 2,4; 2,6: no odor perception, no effects
0.05	2,4/2,6: odor noted immediately upon entering the room; after about 5 min of exposure, 3/6 volunteers experienced a slight "tingling" sensation of the eyes described as lacrimation urge without tears
	2,4: weak odor perception, no eye irritation
	2,6: odor was stronger as compared with the 2,4- isomer
0.075	2,6/2,4: odor became stronger; slight burning of the eyes occurred after 1-6 min, but there was no lacrimation; with deeper breaths, volunteers experienced tickling or a slight stabbing pain in the nose
0.08	2,4: slight conjunctival irritation and tickling of nose
	2,6: eye and nose irritation more severe as compared with same concentration of the 2,4- isomer; effects on throat were perceived as dryness, not scratching sensation
0.10	2,4/2,6: eye and nose irritation became more severe described as resembling a cold (catarrh)
	2,4: more pronounced conjunctival irritation and tickling of nose
	2,6: eye and nose irritation more severe as compared with same concentration of the 2,4- isomer; effects on throat were perceived as dryness, not scratching sensation
0.20	2,4: eye irritation was perceived by 2/5 as stinging and uncomfortable
	2,6: eye and nose irritation more severe as compared with same concentration of the 2,4- isomer; effects on throat were perceived as dryness, not scratching sensation
0.50	2,4/2,6: lacrimation, but eye irritation was still tolerable; one had copious nasal secretion that was associated with "stinging" nasal pain; all had scratchy and burning sensations in the throat, without cough
	2,4: eye irritation was perceived by all as stinging and uncomfortable with lacrimation
	2,6: effects similar to the 2,4- isomer
1.3	2,4/2,6: two individuals were able to remain in the room for 10 min; irritation was intolerable; several hours later, cold-like symptoms with cough persisted

[a]Six healthy male volunteers were exposed to one concentration per day in random order.
Source: data from Henschler et al. 1962.

individuals. This study group is considered a hypersusceptible population and was, therefore, not utilized in setting AEGL values.

Results of early human primary irritation testing with 2,4-TDI were summarized by Zapp (1957) and Wilson and Wilson (1959). Fifty percent of subjects reported the least detectable odor at 0.4 ppm, irritation of the nose and throat occurred at 0.5 ppm, and an appreciable odor was noted at 0.8 ppm. Exposure durations were not given.

2.3. Developmental and Reproductive Toxicity

No information was found regarding the potential developmental or reproductive toxicity of TDI in humans.

2.4. Genotoxicity

No information was found regarding the potential genotoxicity of TDI in humans.

2.5. Carcinogenicity

No information was found regarding the potential carcinogenicity of TDI in humans.

2.6. Summary

Fatalities have been reported following accidental exposures to high concentrations of TDI under unusual circumstances. Human responses to TDI were summarized by Woolrich (1982) from data on worker exposures, case reports, and experimental single exposure studies. Pulmonary effects after TDI inhalation may be either a direct irritant response or the result of an immunologic sensitivity that develops over time. Generally, exposure at ≤0.02 ppm does not elicit a response; however, asthmatic subjects may develop minor irritation and subclinical increases in R_{aw} at that concentration. At concentrations between 0.02 ppm and 0.1 ppm, a portion of the population may develop sensitivity with prolonged exposure. Exposure at >0.1 ppm causes irritation of the respiratory tract, and the severity is dependent on the concentration (Woolrich 1982). It is important to note that the odor threshold of 0.05 ppm (Henschler et al. 1962) is approximately the same concentration that causes slight eye irritation. However, an older study reported 0.40 ppm as the least detectable odor in 50% of subjects (Wilson and Wilson 1959).

3. ANIMAL TOXICITY DATA

3.1. Acute Lethality

3.1.1. Guinea Pigs

A 4-h LC_{50} for the guinea pig was calculated to be 12.7 ppm. Exposures were by whole body and concentrations were measured at 0.1, 1.0, 2, 5, 10, 20, or 34 ppm. A total of 76 animals were used in the experiment (gender and number of animals not stated). Animals exhibited concentration-dependent signs of toxicity, such as mouth-breathing, lacrimation, profuse salivation, and restlessness, during exposure. At concentrations above 5 ppm, mouth-breathing was observed after 1 h of exposure. Histopathologic examinations of the respiratory tracts of five animals per group per time point revealed focal coagulation necrosis and desquamation of the superficial epithelial lining of the trachea and major bronchi. The degree of injury and subsequent repair was dependent on exposure concentration. Inflammation cleared by day 7 post-exposure in the 2-ppm group. Advanced bronchiolitis fibrosia obliterans and bronchopneumonia were evident at the higher concentrations. The specific TDI isomers studied were not identified (Duncan et al. 1962).

3.1.2. Rabbits

A 4-h LC_{50} for the rabbit was estimated to be 11 ppm. Exposures were by whole body and concentrations were measured at 0.1, 1.0, 2, 5, 10, 20, or 34 ppm. A total of 41 animals were used in the experiment (gender and numbers of animals not stated). Animals exhibited concentration-dependent signs of toxicity during exposure such as mouth-breathing, lacrimation, salivation, and restlessness. At concentrations above 5 ppm, mouth-breathing was observed after 1 h of exposure. Histopathologic examinations of the respiratory tracts of two animals per group per time point revealed focal coagulation necrosis and desquamation of the superficial epithelial lining of the trachea and major bronchi. The degree of injury and subsequent repair was dependent on exposure concentration. Inflammation cleared by day 7 post-exposure in the 2-ppm group. Advanced bronchiolitis fibrosia obliterans and bronchopneumonia were evident at the higher concentrations. The specific TDI isomers studied were not identified (Duncan et al. 1962).

3.1.3. Rats

A 4-h LC_{50} for the rat was calculated to be 13.9 ppm. Exposures were by whole body and concentrations were measured at 0.1, 1.0, 2, 5, 10, 20, or 34 ppm. A total of 86 animals were used in the experiment (gender and numbers of animals not stated). Animals exhibited concentration-dependent signs of toxicity during exposure such as mouth-breathing, lacrimation, salivation, and restlessness. At concentrations above 5 ppm, mouth-breathing was observed after 1 h of exposure. Among surviving animals, histopathologic examination of the respiratory tracts of five animals per group per time point revealed focal coagulation necrosis and desquamation of the superficial epithelial lining of the trachea and major bronchi. The degree of injury and subsequent repair was dependent on exposure concentration. Inflammation cleared by day 7 post-exposure in the 2-ppm group, but advanced bronchiolitis fibrosia obliterans and bronchopneumonia were observed at higher concentrations. The TDI isomer mix was not specified (Duncan et al. 1962).

In contrast to the above report, Kimmerle (1976) calculated 4-h LC_{50}s for male and female Wistar II rats (n = 10/gender) to be 49.2 ppm and 50.6 ppm, respectively. Labored breathing was noted during the whole-body exposure, and lung edema and pneumonia were observed at necropsy. The TDI used was identified only by the trade name T 65. The results are higher than the 4-h LC_{50}s reported by Duncan et al. (1962).

A 1-h LC_{50} for Alderley Park male and female albino rats (n = 4/gender; whole-body exposure) was reported at 66 ppm for 2,4- and 2,6-TDI (80:20). No differences were observed between males and females, and most deaths occurred by 36 h post-exposure. At necropsy, all animals showed hemorrhagic edema in the lungs (Horspool and Doe 1977).

Albino rats were exposed by whole body for 6 h to analyzed concentrations of TDI at 2, 4, or 13.5 ppm (mixed isomer, ratio not specified). At both the middle and high concentrations, three of six rats died, those deaths occurring by post-exposure days 7 and 15, respectively. No deaths occurred at 2 ppm. Ocular and nasal irritation and labored breathing were observed at all concentrations. Deaths resulted from severe pulmonary hemorrhage, emphysema, and pneumonia (Wazeter 1964a).

A calculated concentration of 2,4-TDI at 600 ppm for 6 h resulted in pulmonary congestion and edema and was lethal to rats (Zapp 1957).

3.1.4. Mice

A 4-h LC_{50} for the mouse was calculated to be 9.7 ppm. Exposures were by whole body and concentrations were measured at 0.1, 1.0, 2, 5, 10, 20, or 34 ppm. A total of 120 animals were used in the experiment (gender and numbers of animals not stated). Animals exhibited concentration-dependent signs of toxicity during exposure such as mouth-breathing, lacrimation, salivation, and restlessness. At concentrations above 5 ppm, mouth-breathing was observed after 1 h of exposure. Histopathologic examination of the respiratory tracts of five animals per group per time point revealed focal coagulation necrosis and desquamation of the superficial epithelial lining of the trachea and major bronchi. The degree of injury and subsequent repair was dependent on exposure concentration. Inflammation cleared by day 7 post-exposure in the 2-ppm group. Advanced bronchiolitis fibrosia obliterans was evident at the higher concentrations. The specific TDI isomers studied were not identified (Duncan et al. 1962).

3.2. Nonlethal Toxicity

3.2.1. Dogs

As part of a subchronic study, four male dogs were exposed 35-37 times over a period of 4 mo to analytical concentrations of 2,4-TDI averaging 1.5 ppm. Daily exposures were limited to 30 min to 2 h because of the resulting lacrimation, coughing, restlessness, and profuse frothy white secretions from their mouths. The onset of those clinical signs was not specifically noted except that they "continued throughout the entire course of exposure." No deaths were reported (Zapp 1957).

3.2.2. Guinea Pigs

Female albino Dunkin-Hartley guinea pigs (n = 10) were exposed by whole body to 2,4- and 2,6-TDI (80:20) at 3 ppm for 1 h. Clinical signs of toxicity during exposure, if there were any, were not reported. Increased bronchial responsiveness to acetylcholine was evident within 30 min after exposure and lasted up to 48 h. Bronchoalveolar lavage revealed an influx of neutrophils beginning at 1 h post-exposure and lasting approximately 48 h. In related experiments, following continuous exposure at 0.08 ppm for 48 h or at 0.046 ppm for 1 week (wk), bronchial hyper-responsiveness occurred in the absence of neutrophil influx (Gagnaire et al. 1996).

As part of an immunologic study on sensitization to TDI (isomer not specified), female English smooth-haired guinea pigs (n = 8-16) were exposed head-only for 5 d, 3 h/d at concentrations ranging from 0.12 ppm to 7.60 ppm. Sensory irritation was measured as decreased respiratory rate. During a single 3-h exposure, the decrease in respiratory rate was concentration dependent from 0.12 ppm to 0.93 ppm, with maximal response during the first 2 h. At higher concentrations, the maximal respiratory rate decrease (approximately 60%) occurred within the first 30 min (Karol et al. 1980; Karol 1983).

Albino guinea pigs (gender and numbers of animals not stated) were exposed to TDI (method of exposure not stated, mixed isomers assumed) at concentrations ranging from 0.02 ppm to 0.5 ppm for three exposures lasting 6 h each. During a single exposure, concentrations up to 0.05 ppm did not affect the breathing rate. However, at concentrations of ≥0.18 ppm, the breathing rate dropped by 50% after the first 40 min of exposure and by an additional 10% over the next 3.5 h (Stevens and Palmer 1970). It should be noted that the authors did not state whether that pattern developed at the first exposure or whether similar or more severe results occurred with subsequent exposures.

Female English smooth-haired guinea pigs (n = 8) were exposed head-only to 2,4- and 2,6-TDI (80:20) at 1.4 ppm for 3 h/d for 5 consecutive days. Body-weight loss occurred during the exposure period and body weights remained lower than the unexposed controls until termination on day 50 post-exposure. The ventilatory response of exposed animals to 10% CO_2, as measured by pressure change (ΔP), was diminished by 30-50% on day 5 of exposure but gradually recovered during the following 40 d. At sacrifice, exposed animals had multifocal interstitial inflammation of the lungs. In contrast, no adverse effects were observed in animals exposed at 0.02 ppm 6 h/d, 4 d/wk for 70 d (Wong et al. 1985).

Respiratory sensitization was studied in groups of 11-12 male Hartley guinea pigs (Warren 1994a,b). Animals were sensitized with either room air, 2,4-TDI, or 2,6-TDI by nose-only exposure for 3 h/d for 5 d at analytical TDI concentrations of 1.29-1.4 ppm. Clinical signs of toxicity from sensitization exposure included rapid breathing (2,6-TDI), ataxia (2,4-TDI), tremors (2,4- and 2,6-TDI), and death of two animals (one with each isomer). The animals were then challenged by whole-body exposure for 1 h three times at 1 wk intervals with 2,4-TDI or 2,6-TDI at concentrations ranging from 18 ppb to 46 ppb. Challenge concentrations were low enough to avoid sensory irritation and to avoid interfering with a hypersensitivity reaction. Increased respiratory rate was taken as an indicator of hypersensitivity response. No immediate- or delayed-type hypersensitivity reactions were observed in the sham-sensitized animals. On the other hand, both delayed- and immediate-type hypersensitivity reactions occurred in all sensitized groups. Furthermore, the data showed that both 2,4- and 2,6-TDI caused sensitization in the guinea pigs, and that either isomer elicited a hypersensitive reaction regardless of the isomer used for sensitization. However, at necropsy, only the animals given sensitization and challenge exposures to 2,6-TDI showed an increase in red zones in the lungs, suggesting that 2,6-TDI is more irritating than 2,4-TDI.

3.2.3. Hamsters

Male and female Syrian hamsters (n = 5/gender) were exposed to TDI at 0.1 ppm or 0.3 ppm for 6 h/d, 5 d/wk for 4 wk. The TDI isomer or mixture and the method of exposure were not stated. At the high dose, both genders had focal hyperplasia accompanied by slight inflammation of the nasal turbinates and peribronchiolar aggregates of primary mononuclear cells in the lung. Female hamsters also had slight inflammation of the respiratory epithelium of the nasal turbinates from exposure at 0.1 ppm (Kociba et al. 1979).

3.2.4. Rats

Male Sprague-Dawley rats (n = 4) were exposed by head only to 2,4-TDI for 3 h at 0.29, 0.88, 1.41, or 3.20 ppm. Lacrimation and rhinorrhea were observed at all concentrations, labored breathing occurred at ≥1.41 ppm, red swollen conjunctiva were seen at 3.20 ppm, and rales were heard at 0.88 ppm and 3.20 ppm. No mortality occurred. Post-exposure decreases in weight gains occurred in animals exposed at ≥0.88 ppm, but recovery was complete by day 7, except at the highest dose. The respiratory frequency of the rats was concentration-dependent and indicated upper respiratory irritation. The 3-h RD_{50} (concentration which resulted in a 50% decrease in the respiratory rate) was estimated to be 1.37 ppm (Shiotsuka 1987a). In a similar experiment, male Sprague-Dawley rats (n = 4) were exposed head-only for 3 h to a 2,4- and 2,6-TDI mixture (80:20). Concentrations ranged from 0.10 ppm to 1.45 ppm. Transient decreases in weight gain occurred post-exposure at the two highest concentrations, and rales were heard in one animal exposed at 1.45 ppm. The estimated RD_{50} was 2.12 ppm (Shiotsuka 1987b). It should be noted that the estimated RD_{50} for the second study was outside the range of exposure concentrations. Of particular interest, however, was the initial sharp drop in respiratory rate during the first 15 min, followed by a gradual decline during the remainder of the exposure period.

Male Fischer-344 rats (n = 4) exposed head-only to 2,4-TDI at 2 ppm for 4 h appeared lethargic and were not drinking water or eating. However, 12 h post-exposure the animals appeared normal and had resumed eating and drinking (Timchalk et al. 1992). No deaths were reported in rats exposed to a calculated concentration of 2,4-TDI at 60 ppm for 6 h (Zapp 1957).

Albino rats (n = 6) exposed by whole body to TDI (mixed isomer not defined) at 2 ppm for 6 h exhibited ocular and nasal irritation and labored breathing within 2 h of initiation of exposure (Wazeter 1964a). No signs of clinical toxicity were observed at concentrations <1 ppm for 6 h (Wazeter 1964a) or at 0.25 ppm for 8 h (Wazeter 1964b).

3.2.5. Mice

Male Swiss-Webster mice (n=4) were exposed head-only to concentrations of 2,4-TDI ranging from 0.007 ppm to 2.0 ppm for up to 240 min (Sangha and Alarie 1979) or to varying concentrations of 2,6-TDI for 3 h (Weyel et al. 1982). Respiratory irritation was measured as a reduction in respiratory rate. At concentrations of 2,4-TDI above 0.07 ppm, the degree of the response was concentration-dependent with a first maximum reached after 10 min. Following this initial 10-min period, a further decline in respiratory rate was measured gradually for the next 30 min, approximately. The results with 2,6-TDI were described as similar to those with 2,4-TDI indicating respiratory irritation. RD_{50} concentrations at various time points are given in Table 4-5. It is apparent from the development of the response and the $C \times t$ values that respiratory irritation is mainly dependent on concentration and only slightly dependent on duration of exposure. In another series of experiments, those same authors showed that the decrease in respiratory rate was due to irritation of the upper respiratory tract, because exposure by intratracheal instillation failed to result in decreased respiratory rate (Sangha and Alarie 1979). The effect of exposure on respiratory rate also was investigated using concentrations above or below the 1979 TLV (0.02 ppm) for 3 h on each of 5 consecutive days. At exposure concentrations above the 1979 TLV, the level of response was increased and the onset of reaction was faster on each subsequent day. Below the 1979 TLV, no response at all was observed on the first or subsequent days.

3.3. Developmental and Reproductive Toxicity

Pregnant Sprague-Dawley rats (n = 25) were exposed by whole-body inhalation to technical grade TDI (80% 2,4-TDI, 20% 2,6-TDI) at 0.021, 0.120, or 0.480 ppm for 6 h/d on gestation days 6-15. Maternal toxicity at the highest concentration was evident by decreased body weight and weight gain, reduced food consumption, nasal discharge, and audible respiratory distress. Signs of respiratory irritation did not appear until 5 d after treatment began. Fetotoxicity was evinced by delayed ossification of cervical centrum 5 in fetuses from high-concentration litters. No other signs of developmental toxicity were observed (Tyl 1988).

Male and female Sprague-Dawley rats (n = 28/gender) were exposed continuously by whole-body inhalation to technical grade TDI (80% 2,4-TDI, 20% 2,6-TDI) at 0.02, 0.08, or 0.3 ppm for two generations. Exposure of F_0 and F_1 females was discontinued from gestation day 19 through lactation day 4. Clinical signs of toxicity in the adult animals consisted of nasal discharge in F_0 males and red-tinged fur on the head in F_0 females at 0.3 ppm and perinasal encrustation in F_1 females at 0.08 ppm and 0.3 ppm. Histopathologic examination revealed rhinitis in the nasal turbinates of the F_0 adults and the F_1 females at $\geq$0.08 ppm and in the F_1 males at all dose levels. F_2 pup body weights and weight gains were reduced at 0.08 ppm and 0.3 ppm during lactation. There were no treatment-related effects on

TABLE 4-5 Calculated RD_{50} Values in Mice

Exposure time (min)	RD_{50} (ppm)	$C \times t$
2,4-TDI[a]		
10	0.813	8.13
30	0.498	14.94
60	0.386	11.58
120	0.249	29.88
180	0.199	35.82
240	0.199	47.76
2,6-TDI[b]		
180	0.26	46.8

[a]Data from Sangha and Alarie 1979.
[b]Data from Weyel et al. 1982.

the reproductive parameters of either generation (Tyl and Neeper-Bradley 1989).

3.4. Genotoxicity

No information was found regarding potential genotoxicity of TDI in laboratory animals.

3.5. Chronic Toxicity and Carcinogenicity

Results of oncogenicity bioassays with TDI are conflicting and depend on the route of administration. In an NTP (1986) study, groups of 50 male and female F-344/N rats and $B6C3F_1$ mice were given 2,4- and 2,6-TDI (80:20) by gavage 5 d/wk for 2 y. Doses were 60 mg/kg or 120 mg/kg for female rats and mice, 30 mg/kg or 60 mg/kg for male rats, and 120 mg/kg or 240 mg/kg for male mice. Reduced survival was seen in all treated rats and high-dose male mice. Increased subcutaneous fibromas or fibrosarcomas in male rats, mammary fibroadenomas in female rats, and hemangiomas or hemangiosarcomas and hepatocellular adenomas in female mice were observed. In contrast to the NTP (1986) results, a study commissioned by the International Isocyanate Institute (Loeser 1983; Owen 1983) failed to show any evidence of carcinogenicity in Sprague-Dawley CD rats (n = 104-105/gender) or CD-1 mice (n = 89-90/gender) exposed to 2,4- and 2,6-TDI (80:20) by inhalation at 0.05 ppm or 0.15 ppm 6 h/d, 5 d/wk for approximately 2 y. Histopathologic analyses of the nasal turbinates showed a concentration-related increase in rhinitis in both mice (Loeser 1983) and rats (Owen 1983).

The studies described above have been criticized on technical and toxicologic merit. Corn oil was used as vehicle in the gavage study, even though a precipitate of unknown composition formed with the TDI, and TDI is known to breakdown in corn oil (CMA 1989). In the inhalation study, clinical effects were minimal, indicating that exposure concentrations may have been inadequate, but the histopathology of rhinitis confirms that a maximum tolerated dose was achieved (CMA 1989). Despite the suggested scientific flaws of both studies, the route-specific dependence of carcinogenicity may be due to the formation of toluene diamine (TDA), which is the major metabolite produced following oral exposure, but not inhalation exposure (Timchalk et al. 1992, 1994). TDA has previously been shown to be a carcinogen to rats and mice in chronic feeding studies producing tumors similar to those seen in the oral TDI study (NCI 1979, as cited in Timchalk et al. 1994). From histopathologic evaluation, the upper respiratory tract appears to be the target organ following inhalation exposure to TDI, with the response attributable to local irritation.

EPA (1996) has not classified the carcinogenicity of TDI. Based only on the oral studies and the similarity in the tumor response of mice and rats to TDI and TDA, IARC (1985) classified TDI in Group 2B, sufficient evidence of carcinogenicity in animals but inadequate evidence in humans.

3.6. Summary

Animal data on the toxicity of TDI are summarized in Table 4-6. Results of several animal experiments confirm that TDI is a respiratory tract irritant. That was characterized in studies in rats (Shiotsuka 1987a,b) and mice (Sangha and Alarie 1979) showing initial rapid decreases in respiratory rate followed by continued gradual decline. In a series of LC_{50} experiments with rats, mice, guinea pigs, and rabbits, animals exhibited concentration-dependent signs of toxicity during exposure such as mouth-breathing, lacrimation, salivation, and restlessness. Histopathologic examination of the respiratory tract revealed focal coagulation necrosis and

TABLE 4-6 Summary of Animal Toxicity Data

Species	Duration and End Point	Concentration (ppm)	Isomer	Reference
Rat	4-h LC_{50}	13.9	Unknown	Duncan et al. 1962
Male rat	4-h LC_{50}	49.16	Unknown	Kimmerle 1976
Female rat	4-h LC_{50}	50.56	Unknown	Kimmerle 1976
Male and female rat	1-h LC_{50}	66	2,4- and 2,6- (80:20)	Horspool and Doe 1977
Mouse	4-h LC_{50}	9.7	Unknown	Duncan et al. 1962
Guinea pig	4-h LC_{50}	12.7	Unknown	Duncan et al. 1962
Rabbit	4-h LC_{50}	11	Unknown	Duncan et al. 1962
Dog	30-120 min; coughing, lacrimation, restlessness	1.3	2,4-	Zapp 1957
Rat	3-h RD_{50}	1.37	2,4-	Shiotsuka 1987a
Rat	3-h RD_{50}	2.12	2,4- and 2,6- (80:20)	Shiotsuka 1987b
Rat	6 h; ocular and nasal irritation, labored breathing	2	Mixed, not defined	Wazeter 1964a
	6 h; 3/6 dead	4 and 13.5		
Mouse	10-min RD_{50}	0.813	2,4-	Sangha and Alarie 1979
Mouse	4-h RD_{50}	0.199	2,4-	Sangha and Alarie 1979

desquamation of the superficial epithelium lining the trachea and major bronchi. The degree of injury and subsequent repair was dependent on exposure concentration. Inflammation cleared by day 7 post-exposure in the 2-ppm group, but advanced bronchiolitis fibrosia obliterans was observed at higher concentrations. All species but the mouse developed bronchopneumonia following TDI inhalation (Duncan et al. 1962).

Subchronic or chronic inhalation studies in rats, mice, and hamsters indicate that the nasal turbinates are the primary target organ, and the nasal histopathology can be attributed to irritation.

4. SPECIAL CONSIDERATIONS

4.1. Metabolism and Disposition

In two related studies, healthy men (ages 36-50) were exposed to TDI at concentrations of approximately 3.5, 7, and 9.8 ppb (25, 50, and 70 $\mu g/m^3$) for 4 h at 1 wk intervals (Brorson et al. 1991) or at 5.6 ppb (40 $\mu g/m^3$) for 7.5 h (Skarping et al. 1991). Acetylator phenotype was assessed in the subjects of the Skarping et al. (1991) study. The isomeric composition of the air in the test chamber was 30% 2,4-TDI and 70% 2,6-TDI in the first study and 48% 2,4-TDI and 52% 2,6-TDI in the second study. Plasma concentrations of 2,4- and 2,6-TDA were analyzed over a period of up to 5 wk after the initial exposure. There were concentration- and time-dependent increases in plasma levels of TDA, with 2,6-TDA appearing after exposure at 5.6 ppb or 7.02 ppb (40 $\mu g/m^3$ or 50 $\mu g/m^3$) and 2,4-TDA detectable after exposure at 5.6 ppb or 9.8 ppb (40 $\mu g/m^3$ or 70 $\mu g/m^3$). Similar or slightly higher plasma levels were detected 24 h after exposure. The plasma elimination half-life for the initial rapid phase was calculated to be about 4-5 h for 2,6-TDA and 2-3 h for 2,4-TDA. Half-life for the slower elimination phase was not given. Inhaled doses were calculated from ventilation rates and exposure concentrations and durations. Cumulative urinary excretion of 2,4- and 2,6-TDA directly correlated with the concentrations of 2,4- and 2,6-TDI, respectively, in the test chamber. Over 24-28 h, the cumulative amount of 2,4-TDA excreted in the urine was 8-19% of the estimated inhaled dose of 2,4-TDI, and that of 2,6-TDA was 14-23% of the estimated inhaled dose of 2,6-TDI. No differences were observed between fast and slow acetylators (Brorson et al. 1991; Skarping et al. 1991).

Male Hartley guinea pigs were exposed by whole-body inhalation to ^{14}C-labeled 2,4-TDI for 1 h. The rate of uptake into the bloodstream was linear over a concentration range of 0.004-0.146 ppm, and there was a continued slight increase post-exposure. The level of radioactivity in the bloodstream declined gradually over 72 h but did not show a significant decline over the subsequent 11 d period. Immediately following exposure, most of the radioactivity was distributed to the trachea, and smaller amounts were found in the lung, kidney, heart, spleen, and liver. Elimination was mainly through the urine (Kennedy et al. 1989).

Male Fischer-344 rats were exposed by inhalation to ^{14}C-labeled 2,4-TDI at 2 ppm for 4 h. It was estimated that essentially all of the inhaled TDI was retained by the animal. The half-life for urinary elimination was approximately 20 h. Acid labile conjugates accounted for about 90% of the urinary metabolites, and 10% was acetylated TDA. In contrast, the major urinary metabolite following oral administration was 2,4-TDA (Timchalk et al. 1992, 1994).

In summary, although the systemic uptake of TDI follows linear $C \times t$ kinetics, that relationship does not hold for the onset of signs of toxicity over the same concentration range. For sensory irritation (Sangha and Alarie 1979) and for development of sensitization (Karol 1983; Garabrant and Levine 1994), the response is mainly concentration-dependent.

4.2. Mechanism of Toxicity

Inhaled TDI is corrosive, and the parent material acts as a direct chemical irritant. The degree of irritation appears to be dependent on concentration rather than duration of exposure (Duncan et al. 1962; Sangha and Alarie 1979; Bernstein 1982). In both human and animal studies, an immediate decline in respiratory rate occurs with onset of exposure, followed by a continued, more gradual decline (Baur 1985; Sangha and Alarie 1979; Weyel et al. 1982). Subchronic or chronic inhalation studies in rats, mice, and hamsters indicate that the nasal turbinates are the primary target organ in rodents and that the frank pathology there can be attributed to direct chemical deposition and irritation (Kociba et al. 1979; Loeser 1983; Owen 1983).

It has long been established that repeated inhalation contact with TDI can provoke asthmatic reactions in humans (Zapp 1957). Immediate, late, and dual asthmatic responses have been documented in sensitized individuals (Butcher et al. 1979; Karol 1986). Karol (1986) concluded that although TDI does not cause asthma by a nonspecific irritant effect, concomitant irritation or hyper-reactivity of the airways may produce heightened respiratory tract responsiveness in isocyanate-sensitive individuals. A review of epidemiological studies that reported sensitization rate found that in five of six worker populations the rate of sensitization was between 0 and 1.5% per year. In the sixth worker population, a majority of air samples showed TDI exposures above 20 ppb (142 $\mu g/m^3$) and a rate of sensitization of 5% per year (Garabrant and Levine 1994). Two distinct populations with occupational asthma were identified on the basis of the duration of exposure to TDI before the onset of symptoms. One group developed asthma after an average of 2.4 y, while the other developed asthma after an average of 21.6 y (Di Stefano et al. 1993).

The mechanism by which TDI induces asthmatic symptoms is not entirely known, but it appears to include both immunologic and nonimmunologic mechanisms (Bernstein 1982). Proposed mechanisms include pharmacologic bronchoconstriction, allergic or immunologically mediated bronchoconstriction, and hyper-reactive airways (Karol 1986). In general, isocyanates are reactive substances capable of antigenic activity (Woolrich 1982). Although, the guinea pig has been widely used as a model for TDI-induced asthma, pulmonary hyper-reactivity in guinea pigs only works for sensitized animals challenged with TDI-protein conjugate, whereas sensitized humans react to TDI alone (Karol et al. 1980). Some studies showed detection of IgE antibodies in sensitized individuals, although others found variable or negative results (Karol 1986). IgG antibodies have been detected in both healthy and symptomatic workers (Baur 1985). The inflammatory response of the respiratory tract has been characterized by persistent activation of lymphocytes, chronic expression of certain cytokines (Maestrelli et al. 1995), neutrophilia, eosinophilia (Fabbri et al. 1987), and decreased lymphocyte cAMP levels (Butcher et al. 1979). Individuals that developed asthma after short-term exposure were shown to have a greater number of mast cells in their airway mucosa than individuals that developed asthma after longer-term exposure (Di Stefano et al. 1993). Direct application of TDI in vitro induced the release of tachykinins from sensory nerves in the isolated mouse trachea (Scheerens et al. 1996). The mechanisms behind TDI-induced asthma have been thoroughly reviewed elsewhere (Karol 1986), but several major areas are discussed below.

Guinea pigs were sensitized by exposure to TDI at concentrations ranging from 0.12 ppm to 10 ppm for 3 h/d for 5 consecutive days (Karol et al. 1980; Karol 1983). All animals exposed at 10 ppm died following exposure on day 3. Twenty-two days later, animals were evaluated for TDI-specific antibodies. No antibodies were detected in animals that inhaled 0.12 ppm. However, 55% of animals exposed at ≥0.36 ppm had serum antibodies. Higher concentrations of TDI resulted in a greater percentage of animals producing antibodies and in higher antibody titers. When challenged, a significant association was found with lung sensitivity (increased respiratory rate) and the presence of circulating antibodies, rather than with the antibody titer (Karol 1983). Increased respiratory rate is a well-documented phenomenon that occurs in immunologically sensitized animals and humans following inhalation of specific antigens. Increased respiration is probably a reflex due to the hypoxia resulting from narrowing of the airway lumen.

In a similar experiment, respiratory sensitization was studied in groups of 11-12 male Hartley guinea pigs (Warren 1994a,b). Animals were sensitized with room air, 2,4-TDI, or 2,6-TDI by exposure at analytical TDI concentrations of 1.29-1.4 ppm for 3 h/d for 5 d. The animals were then challenged with 2,4- or 2,6-TDI protein conjugates at concentrations ranging from 18 to 46 ppb for 1 h three times at 1 wk intervals. Challenge concentrations were low enough to avoid sensory irritation and to avoid interfering with a hypersensitivity reaction. Increases in respiratory rate were taken as indicators of hypersensitivity responses. No immediate- or delayed-type hypersensitivity reactions were observed in the sham-sensitized animals. Both delayed- and immediate-type hypersensitivity reactions occurred in all sensitized groups. Furthermore, the data showed that both 2,4-TDI and 2,6-TDI caused respiratory sensitization in the guinea pigs, and that either isomer elicited a hypersensitive reaction regardless of the isomer used for sensitization.

Both of the above studies used guinea pigs as a model for TDI-induced pulmonary hypersensitivity. But, despite the presence of circulating antibodies demonstrated by Karol (1983), inhalation challenge of animals elicited pulmonary sensitivity only when sensitized animals were challenged with TDI-protein conjugates and not TDI alone.

Another study in guinea pigs showed that dose-response relationships exist for both induction and challenge concentrations for production of TDI sensitization as measured by histamine release and mast cell degranulation (Huang et al. 1993).

4.3. Structure-Activity Relationships

TDI exists in both the 2,4- and 2,6- isomeric forms, which are available commercially as 65:35 or 80:20 mixtures. Most studies with TDI fail to specify the isomer or mixture employed. However, in the studies that did state which isomer was used, there appeared to be little difference in toxicity between the two. In humans, 2,6-TDI was slightly more irritating than 2,4-TDI, but the irritant potential of the 2,6-isomer was similar to that of the mixture (Henschler et al. 1962). Studies by Warren (1994a,b) showed that both 2,4- and 2,6-TDI caused sensitization in the guinea pigs and that either isomer elicited a hypersensitive reaction regardless of the isomer used for sensitization. Animals in that study that were sensitized and challenged with 2,6-TDI developed gross lung lesions (red zones), indicating that the 2,6- isomer is the more irritating. However, the 3-h RD_{50} in mice was approximately 0.2 ppm for both the 2,4- (Sangha and Alarie 1979) and 2,6- (Weyel 1982) isomers.

Kimmerle (1976) found that the LC_{50} for the TDI polymer was about 10 times greater than the LC_{50} for monomeric TDI (designated T 65 with no isomer identification) in rats. The polymers used in that study were Desmodur L 67, Desmodur IL, and Desmodur HL.

Little information was found on cross-reactivity with other isocyanates in individuals sensitized to TDI. Karol (1986) noted that in the workplace, individuals are neither exposed to nor sensitized by monoisocyanates; rather, sensitization is a result of exposure to diisocyanates. However, the monoisocyanates have been successfully used as haptens in detecting antibodies to the corresponding diisocyanate (Karol 1986), and *p*-tolyl isocyanate has been used to detect antibodies in TDI-sensitized individuals (Karol et al. 1980).

TDI is structurally similar to methyl isocyanate (MIC). The databases for these isocyanates are robust and each contains animal and human studies. However, their only consistent similarity is that both are irritants when inhaled, and the available data suggest differing mechanisms of action beyond irritation. TDI is a proven sensitizer, MIC is not. Systemic effects have been well-documented after MIC inhalation exposure but not after TDI exposure. For example, in laboratory animal studies, the fetal and neonatal deaths resulting from inhalation exposure to MIC did not occur following maternal exposure to TDI. Cardiac arrhythmias reported after MIC exposures have not been seen after exposures to TDI. For MIC, systemic effects may occur at concentrations equal to or below those that cause irritation. Therefore, although TDI and MIC are both isocyanates, the end points chosen for the derivation of AEGL values differed for each chemical based on the available data from inhalation exposures.

4.4. Other Relevant Information

4.4.1. Species Variability

Although 4-h LC_{50} values for the rat and mouse do not differ appreciably (13.9 ppm and 9.7 ppm, respectively) (Duncan et al. 1962), the RD_{50}

values for rats (Shiotsuka 1987a,b) are approximately 10 times greater than those for mice (Sangha and Alarie 1979). Animal models have been validated for the mouse (Alarie 1981) and guinea pig (Borm et al. 1990). On the basis of those data, it appears that the mouse is the common laboratory animal most sensitive to the irritating effects of TDI.

The guinea pig has been studied extensively as a model for TDI-induced asthma. However, pulmonary hyper-reactivity in guinea pigs only develops in sensitized animals challenged with TDI-protein conjugate (Karol et al 1980). In contrast, sensitized humans react to TDI alone.

4.4.2. Sensitive Subpopulations

As discussed in Section 4.2, TDI produces asthmatic reactions in sensitized individuals. Rates of sensitization in workers were found to range from 0 to 5%, and the highest rate correlated to TDI exposures above 20 ppb (Garabrant and Levine 1994). The mechanism by which TDI induces asthma is not known, nor are data available to quantify the rate of sensitization in the general population. The presence of circulating antibodies has not proved to be a reliable indicator of sensitization or symptomology (Karol 1986). Therefore, at the AEGL levels there may be individuals that have a strong reaction to TDI, and those individuals may not be protected.

4.4.3. Unique Physicochemical Properties

Several physicochemical properties of TDI minimize the opportunity for acute inhalation exposure to high concentrations. The low vapor pressure (0.01 mm Hg at 20°C) corresponds to a saturated atmospheric concentration of 14.9 ppm (Horspool and Doe 1977). Temperature must be increased before higher concentrations are possible. Also, TDI readily reacts with water vapor resulting in a "fall-out" of reaction product that is probably TDI-urea (Zapp 1957; Wazeter 1964a; Horspool and Doe 1977). Deposition and reaction with moisture can act to reduce the atmospheric concentration of TDI. These phenomena are responsible for large differences in theoretical vs analytical exposure concentrations (Wazeter 1964a) and probably explain the lack of effects reported by Zapp (1957) at concentrations that resulted in clear effects, including death, in other studies (Wazeter 1964a,b; Duncan et al. 1962; Sangha and Alarie 1979).

4.4.4. Concentration-Exposure Duration Relationship

The concentration-exposure duration relationship for an irritant gas such as TDI can be described by the equation $C^n \times t = k$, where the exponent n ranges from 0.8 to 3.5 (ten Berge et al. 1986). In the absence of a chemical-specific, empirically derived exponent, a default value of $n = 1$ can be used when extrapolating to longer time points, and a default value of $n = 3$ can be used when extrapolating to shorter time points. This method will yield the most conservative AEGL estimates.

5. DATA ANALYSIS AND AEGL-1

AEGL-1 is the airborne concentration (expressed as ppm or mg/m^3) of a substance above which it is predicted that the general population, including susceptible individuals, could experience notable discomfort, irritation, or certain asymptomatic, nonsensory effects. However, the effects are not disabling and are transient and reversible upon cessation of exposure.

5.1. Summary of Human Data Relevant to AEGL-1

The human data most relevant to AEGL-1 are those of the Baur (1985) study in which both asthmatics and healthy volunteers were exposed to controlled concentrations of 2,4- and 2,6-TDI (80:20). Asthmatic subjects were exposed to TDI at 0.01 ppm for 1 h. Then, after a rest of 45 min, they were exposed at 0.02 ppm for 1 h. A referent group of nonasthmatic subjects was exposed to TDI at 0.02 ppm for 2 h. Although no statistically significant differences in lung function parameters were observed among asthmatic subjects during or after exposure, bronchial obstruction was indicated in several subjects. Individually, no decrease in FEV_1 of more than 20% was observed. The magnitude of airway resistance was not considered clinically significant for the asthmatic subjects, indicating that those effects fall within the definition of AEGL-1. In the healthy referent group, there was a significant increase in airway resistance immediately after and 30 min after the beginning of exposure, but none of the subjects developed bronchial obstruction. Both groups reported eye and throat irritation, cough, chest tightness, rhinitis, dyspnea, and/or headache, but time to onset of symptoms was not given.

Similar symptoms were reported among spray-foam workers exposed to average isocyanate (isomer not identified) concentrations of up to 0.043 ppm for as long as 7.4 h (Hosein and Farkas 1981). Symptoms of exposure were reported when workplace air concentrations exceeded 0.03 ppm (Hama 1957). Healthy subjects tolerated approximately 0.01 ppm for 4 h with no adverse effects (Brorson et al. 1991).

5.2. Summary of Animal Data Relevant to AEGL-1

None of the available animal data was relevant to derivation of AEGL-1.

5.3. Derivation of AEGL-1

The data of Baur (1985) were used for derivation of AEGL-1 values. Asthmatic individuals tolerated exposure at 0.01 ppm for 1 h, and then, after a rest, 0.02 ppm for another hour. Because the time to onset of symptoms was not identified, it is assumed that the effects began immediately upon TDI exposure. This assumption is supported by the fact that significant differences in lung function occurred in the healthy population immediately after and 30 min after initiation of exposure, but resolved with longer duration of exposure. There was also no indication whether the effects were worse in asthmatic subjects at 0.01 ppm or at 0.02 ppm. Therefore, the 0.02-ppm concentration was identified as the basis for the 10-min, 30-min, and 1-h AEGL-1 values, and the 0.01-ppm concentration was identified as the 4- and 8-h AEGL-1s. Extrapolations across time were not performed. Because the asthmatic subjects tolerated 0.02 ppm for 1 h after pre-exposure at 0.01 ppm, it is assumed that the asthmatic population could tolerate the lower concentration for a longer duration. However, it is recognized that individuals with pre-existing allergic sensitization to TDI might not be protected at those concentrations and might experience airway reactivity with symptoms characteristic of an asthmatic attack, such as coughing, wheezing, chest tightness, and difficulty breathing. It should also be noted that AEGL-1 values are below any reported odor threshold concentrations (Henschler et al. 1962; Wilson and Wilson 1959). AEGL-1 values are presented in Table 4-7.

The AEGL-1 values are considered protective of public health as defined under AEGL-1. Asthmatic subjects were studied, making the use of

TABLE 4-7 AEGL-1 Values for Toluene 2,4- and 2,6-Diisocyanate (ppm [mg/m^3])

10 min	30 min	1 h	4 h	8 h
0.02 (0.14)	0.02 (0.14)	0.02 (0.14)	0.01 (0.07)	0.01 (0.07)

uncertainty factors unnecessary because asthmatic people are considered a sensitive subpopulation. The 0.01-ppm exposure concentration for the longer time points is reasonable because data suggest that the adverse health effects of inhaled TDI are more concentration-dependent than duration-dependent. Controlled inhalation at 0.02 ppm was tolerated by asthmatic subjects for 1 h. For comparison, the spray foam applicators in the Hosein and Farkas (1981) study tolerated up to 4 times the AEGL-1 values (0.04 ppm) for up to 7.5 h with reports of eye irritation only. Assuming that the applicators in the Hosein and Farkas (1981) study were healthy adults (i.e., nonasthmatic), and assuming that the isocyanates measured were TDI, minimal effects would be expected in normal individuals at the AEGL-1 concentrations. Also, healthy subjects tolerated approximately 0.01 ppm for 4 h with no adverse effects (Brorson et al. 1991). A slightly higher concentration of 0.03 ppm resulted in symptoms in 100% of workers at a manufacturing plant (Hama 1957). The AEGL-1 single-exposure values are below the concentrations expected to cause sensitization with repeated long-term exposure (Garabrant and Levine 1994). Figure 4-1 is a plot of the derived AEGLs and all of the human and animal data on TDI.

6. DATA ANALYSIS AND AEGL-2

AEGL-2 is the airborne concentration (expressed as ppm or mg/m^3) of a substance above which it is predicted that the general population, including susceptible individuals, could experience irreversible or other serious, long-lasting adverse health effects or an impaired ability to escape.

6.1. Summary of Human Data Relevant to AEGL-2

The most appropriate human data for use in derivation of AEGL-2 values are those of Henschler et al. (1962). Human subjects were exposed to analytical concentrations of 2,4- and 2,6-, 2,4-, or 2,6-TDI ranging from 0.01 to 1.3 ppm for 30 min. At 0.5 ppm, volunteers experienced ocular

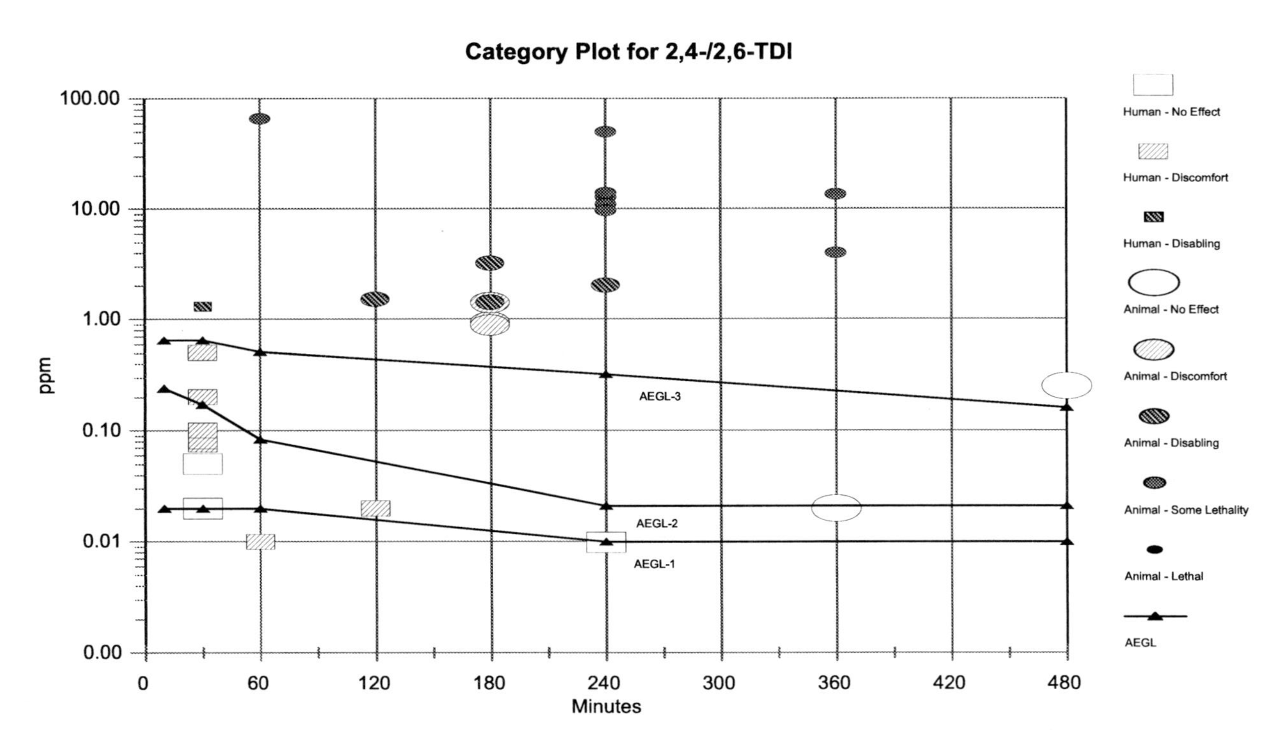

FIGURE 4-1 Toxicity data and AEGL values for toluene 2,4- and 2,6-diisocyanate. Toxicity data include both human and animal studies.

irritation with lacrimation and throat irritation in the absence of cough. Irritation was intolerable at the next higher concentration tested (1.3 ppm), forcing subjects to leave the room after only 10 min; cough persisted for several hours.

6.2. Summary of Animal Data Relevant to AEGL-2

Mouse and rat RD_{50} values were considered for calculation of AEGL-2 values. Decreased respiratory rate in the mouse model has been shown to correspond with sensory irritation in humans. When an irritant such as TDI enters the nasal mucosa, the trigeminal nerve endings are stimulated, resulting in an inhibition of respiration (Alarie 1981). The 10-min and 1-h RD_{50}s for TDI in male Swiss-Webster mice are 0.8 ppm and 0.39 ppm, respectively (Sangha and Alarie 1979). The 3-h RD_{50} in male Sprague-Dawley rats ranged from 1.37 ppm to 2.12 ppm (Shiotsuka 1987a,b). In those experiments, there was an initial sharp drop in respiratory rate during the first 15 min followed by a gradual decline during the remainder of the exposure period. This effect is indicative of concentration-dependent irritation. Fischer-344 rats exposed at 2 ppm for 4 h appeared lethargic and were not drinking water or eating. However, 12 h post-exposure, the animals appeared normal and had resumed eating and drinking (Timchalk et al. 1992). In the series of experiments by Duncan et al. (1962) that exposed four species of laboratory animals at 2 ppm, clearing of the inflammation and respiratory tract injury was apparent by day 7 post-exposure.

6.3. Derivation of AEGL-2

Because rigorous human data are available, they were used to calculate the AEGL-2. Exposure at 0.5 ppm for 30 min resulted in eye and throat irritation and lacrimation. A higher concentration was intolerable after 10 min. Although the extent of lacrimation at 0.5 ppm was not described, any amount could impair the ability to escape. Therefore, that is probably close to a NOAEL for AEGL-2. In addition, the ocular and respiratory tract irritation associated with TDI exposure appears to be more concentration-dependent than duration-dependent. However, exposure for longer periods can result in excessive fluid accumulation in the respiratory tract, which could lead to more severe consequences than those defined under AEGL-2.

Extrapolations across time were made using the equation $C^n \times t = k$, where n ranges from 0.8 to 3.5 (ten Berge et al. 1986). In the absence of an empirically derived, chemical-specific exponent, scaling was performed using $n = 3$ for extrapolating to the 10-min time point and $n = 1$ for the 1- and 4-h time points. The 4-h value was used for the 8-h because extrapolation to 8 h resulted in a concentration similar to one that caused mild effects in polyurethane foam sprayers exposed for >7 h (Hosein and Farkas 1981) and in manufacturing workers on 8-h shifts (Hama 1957). An intraspecies uncertainty factor (UF) of 3 was applied to account for sensitive individuals; use of a greater UF would result in values below those supported by the human data. The values for AEGL-2 are presented in Table 4-8.

An intraspecies UF of 3 has been used before in establishing AEGL values for chemicals that are rapidly acting respiratory irritants. Although some individuals with pre-existing bronchial hyper-reactivity have been shown to respond to TDI with nonpathological bronchial obstruction (4/15), no significant differences were observed in lung function parameters. Also, complaints of respiratory irritation occurred in both asthmatic subjects and healthy controls (Baur 1985).

Borm et al. (1990) used animal data for various toxic end points resulting from TDI exposure to calculate exposure levels for humans. The end points included were respiratory irritation, sensitization, airway hyper-responsiveness, and gradual loss of pulmonary function. The authors found that use of respiratory irritation resulted in the most conservative estimates for protection of human health (Borm et al. 1990). Using one-tenth of the mouse or rat RD_{50}, a measure of respiratory irritation, for calculation of AEGL-2, and applying an intraspecies UF of 3 results in values that are below concentrations shown to affect humans (Hosein and Farkas 1981; Henschler et al. 1962). However, similar results to the AEGL-2 values are obtained by starting with a 4-h exposure at 2 ppm. At that exposure regimen, clearing of respiratory tract lesions was observed in four laboratory species (Duncan et al. 1962). Therefore, the animal data strongly support the AEGL-2 values derived from human experiments.

7. DATA ANALYSIS AND AEGL-3

AEGL-3 is the airborne concentration (expressed as ppm or mg/m^3) of a substance above which it is predicted that the general population, including susceptible individuals, could experience life-threatening health effects or death.

TABLE 4-8 AEGL-2 Values for Toluene 2,4- and 2,6-Diisocyanate (ppm [mg/m^3])

10 min	30 min	1 h	4 h	8 h
0.24 (1.71)	0.17 (1.21)	0.083 (0.59)	0.021 (0.15)	0.021 (0.15)

7.1. Summary of Human Data Relevant to AEGL-3

No reliable human data were available for derivation of AEGL-3 values. Reported human fatalities occurred under unusual circumstances, and exposure concentrations were not measured. Acute exposure reports emphasize that the respiratory tract is the primary target, and pulmonary edema develops subsequent to the irritation brought on by the corrosive properties of TDI.

7.2. Summary of Animal Data Relevant to AEGL-3

On the basis of LC_{50} values, the mouse is the species most sensitive to the effects of TDI. The 4-h LC_{50} for the mouse was 9.7 ppm. Death was preceded by severe pathology in the respiratory tract (Duncan et al. 1962). Mouse RD_{50} values are considered equivalent to AEGL-3 values for humans (Alarie 1981). The 10-min and 1-h RD_{50}s of TDI in male Swiss-Webster mice are 0.8 ppm and 0.39 ppm, respectively (Sangha and Alarie 1979).

7.3. Derivation of AEGL-3

The 4-h mouse LC_{50} of 9.7 ppm (Duncan et al. 1962) was divided by 3 to estimate a threshold of lethality from the regression plot. The LC_{50} probit plot from Duncan et al. (1962) is shown in Appendix A. Extension of the regression line for the mouse data to the x-intercept shows that a concentration at approximately 4 ppm would result in 1% lethality. Therefore, one-third of the LC_{50} is considered to be a reasonable estimate of the threshold for lethality (NRC 2001).

The estimated 4-h lethality threshold, 3.23 ppm, was used to extrapolate to the 30-min and 1- and 8-h AEGL-3 time points. Values were scaled using the equation $C^n \times t = k$, where n ranges from 0.8 to 3.5 (ten Berge et al. 1986). In the absence of an empirically derived, chemical-specific exponent, scaling was performed using $n = 3$ for extrapolating to the 30-min and 1-h time points and $n = 1$ for the 8-h time point. A total UF of 10 was applied, which includes 3 to account for sensitive individuals and 3 for interspecies extrapolation (use of a greater UF would result in values similar to concentrations that produced mild irritation in human inhalation studies). According to Section 2.7 of the standard operating procedures for the derivation of AEGLs (NRC 2001), 10-min values are not to be scaled from an experimental exposure time of ≥4 h. Therefore, the 30-min AEGL-3 value was adopted as the 10-min value. The values for AEGL-3 are given in Table 4-9.

Individuals already sensitized to TDI may exist in the general population. No data are available to quantify or estimate the rate of sensitization. At the AEGL-3 levels, individuals who have a stronger reaction to TDI might not be protected from severe effects.

Using the mouse RD_{50}, a measure of respiratory irritation, to calculate AEGL-3 and applying an intraspecies UF of 3 results in values that are similar to concentrations shown to affect humans in controlled experimental studies (Henschler et al. 1962).

8. SUMMARY OF AEGLS

8.1. AEGL Values and Toxicity End Points

The derived AEGLs for various levels of effects and durations of exposure are summarized in Table 4-10. AEGL-1 and AEGL-2 were based on sensory irritation in humans. The basis for AEGL-3 was a calculated 4-h LC_{50} in the mouse. Presensitized individuals might exist in the general population, but the rate of TDI sensitization cannot be predicted. If the rate of sensitization in the general population were quantifiable, the committee might have considered a different approach to derivation of AEGL values.

TABLE 4-9 AEGL-3 Values for Toluene 2,4- and 2,6-Diisocyanate (ppm [mg/m^3])

10 min	30 min	1 h	4 h	8 h
0.65 (4.63)	0.65 (4.63)	0.51 (3.63)	0.32 (2.28)	0.16 (1.14)

TABLE 4-10 Summary of AEGL Values (ppm [mg/m³])

	Exposure Duration				
Classification	10 min	30 min	1 h	4 h	8 h
AEGL-1 (Nondisabling)	0.02 (0.14)	0.02 (0.14)	0.02 (0.14)	0.01 (0.07)	0.01 (0.07)
AEGL-2 (Disabling)	0.24 (1.71)	0.17 (1.21)	0.083 (0.59)	0.021 (0.15)	0.021 (0.15)
AEGL-3 (Lethal)	0.65 (4.63)	0.65 (4.63)	0.51 (3.63)	0.32 (2.28)	0.16 (1.14)

At each of the AEGL levels, individuals who have a strong reaction to TDI might not be protected within the definition of effects for each level.

8.2. Comparison with Other Standards and Criteria

Existing guideline exposure levels for TDI are listed in Table 4-11. NIOSH has not set exposure limits for 2,4-TDI but recommends limiting exposure to the lowest feasible concentration (NIOSH 1997). The OSHA ceiling limit (concentration that should not be exceeded at any time) for 2,4-TDI is 0.02 ppm (OSHA 1995).

The IDLH is based on acute inhalation toxicity data in animals, but was not based on data obtained from the exposures of humans or animals sensitized to TDI (NIOSH 1996). Four-hour LC_{50} values in four laboratory animal species ranged from 9.7 ppm to 13.9 ppm (Duncan et al. 1962). To calculate the IDLH, these LC_{50}s were adjusted to 30 min by multiplying by a correction factor of 2. The adjusted values were divided by a UF of 10 to yield derived values of 1.9-2.8 ppm. Therefore, the IDLH was set at 2.5 ppm (NIOSH 1996). Those same data were used in derivation of AEGL-3; however, the resulting 30-min AEGL-3 is approximately one-third of the IDLH because an estimation of the threshold for lethality was obtained by dividing the mouse LC_{50} by 3.

ACGIH (2001) classifies the chemical as a sensitizer, which refers to the potential for an agent to produce sensitization. The sensitizer notation does not imply that sensitization is the critical effect on which the Threshold Limit Value (TLV) is based, nor does it imply that the effect is the sole basis for the TLV (ACGIH 2001). The 8-h TLV of 0.005 ppm is intended to both protect against possible sensitization in workers and reduce the opportunity for accidental TDI exposure.

TABLE 4-11 Extant Standards and Guidelines for Toluene 2,4- and 2,6-Diisocyanate (ppm)

	Exposure Duration				
Guideline	10 min	30 min	1 h	4 h	8 h
AEGL-1	0.02	0.02	0.02	0.01	0.01
AEGL-2	0.24	0.17	0.083	0.021	0.021
AEGL-3	0.65	0.65	0.51	0.32	0.16
PEL-TWA (OSHA)[a]					0.02 (C)
IDLH (NIOSH)[b]		2.5			
TLV-TWA (ACGIH)[c]					0.005 (SEN)
TLV-STEL (ACGIH)[d]	0.02 (SEN)				
MAC (The Netherlands)[e]	0.02 (15-min)				0.005

[a]OSHA PEL-TWA (permissible exposure limit-time-weighted average of the Occupational Health and Safety Administration) (29 CFR § 1910.1000). The PEL-TWA is defined analogous to the ACGIH TLV-TWA but is for exposures of no more than 10 h/d, 40 h/wk. (C) denotes a ceiling limit.

[b]IDLH (immediately dangerous to life and health of the National Institute of Occupational Safety and Health) (NIOSH 1996). The IDLH represents the maximum concentration from which one could escape within 30 min without any escape-impairing symptoms or any irreversible health effects. The IDLH for TDI is based on acute inhalation toxicity data in animals, but is not based on data obtained from the exposures of individuals or animals already sensitized to TDI.

[c]ACGIH TLV-TWA (Threshold Limit Value–time-weighted average of the American Conference of Governmental Industrial Hygienists) (ACGIH 1996, 2001). The TLV-TWA is the time-weighted average concentration for a normal 8-h work day and a 40-h work week to which nearly all workers may be repeatedly exposed, day after day, without adverse effect. SEN notation refers to the potential for an agent to produce sensitization.

[d]ACGIH TLV-STEL (Threshold Limit Value–short-term exposure limit) (ACGIH 2001). The TLV-STEL is defined as a 15-min TWA exposure that should not be exceeded at any time during the work day even if the 8-h TWA is within the TLV-TWA. Exposures above the TLV-TWA up to the STEL should not be longer than 15 min and should not occur more than 4 times per day. There should be at least 60 min between successive exposures in that range. SEN notation refers to the potential for an agent to produce sensitization.

[e]MAC (Maximaal Aanvaaarde Concentratie [Maximal Accepted Concentration]) (SDU Uitgevers [under the auspices of the Ministry of Social Affairs and Employment], The Hague, The Netherlands 2000). The MAC is defined analogous to the ACGIH TLV-TWA.

ERPG values for TDI were under consideration as of year 2000 but had not been derived by 2002. The German Research Council (2000) has not recommended a current MAK but lists the chemical as an airway sensitizer.

8.3. Data Adequacy and Research Needs

Limited quantitative data in humans were available for use in deriving AEGLs. Experimental studies in humans included one that used both asthmatic subjects and healthy subjects and another that reported a concentration-response assessment. However, those are the only human studies available. Generally, very low concentrations of TDI were reported in occupational studies. Animal data have shown concentration-dependent effects, including irritation and histologic lesions of the respiratory tract and lethality. Because the nonlethal and lethal effects in humans and animals are qualitatively similar, the animal data were considered relevant and appropriate for developing AEGL values as described in the standing operating procedures of the National Advisory Committee for AEGLs (NRC 2001).

The most notable data deficiencies were the absence of quantitative human exposure data, the absence of a well-defined exposure-response curve for the toxic effects in animals, a lack of understanding of individual variability in the toxic response to TDI, and a lack of information on the extent of cross-reactivity between isocyanates.

Critical research needs include defining thresholds for effects and how those thresholds might vary with exposure concentration and duration. Such data would be valuable for affirming the AEGL values. In addition, a scientifically verifiable estimate of the number of individuals in the general population who are presensitized to TDI would be instrumental in reducing uncertainties in quantitative health risk issues.

9. REFERENCES

ACGIH (American Conference of Governmental Industrial Hygienists). 1996. Toluene-2,4-Diisocyanate. Pp. 1581-1589 in Documentation of the Threshold

Limit Values and Biological Exposure Indicies, 6th Ed. Cincinnati, OH: ACGIH.

ACGIH (American Conference of Governmental Industrial Hygienists). 2001. Threshold Limit Values (TLVs) for Chemical Substances and Physical Agents and Biological Exposure Indicies (BEIs). Cincinnati, OH: ACGIH. Pp. 57.

Alarie, Y. 1981. Dose-response analysis in animal studies: Prediction of human responses. Environ. Health Perspect. 42:9-13.

Baur, X. 1985. Isocyanate hypersensitivity. Final report to the International Isocyanate Institute. III File No. 10349; III Project E-AB-19.

Bernstein, I.L. 1982. Isocyanate-induced pulmonary diseases: A current perspective. J. Allergy Clin. Immunol. 70:24-31.

Borm, P.J.A., T.H.J.M. Jorna, and P.T. Henderson. 1990. Setting acceptable exposure limits for toluene diisocyanate on the basis of different airway effects observed in animals. Reg. Toxicol. Pharmacol. 12:53-63.

Brorson, T., G. Skarping, and S. Carsten. 1991. Biological monitoring of isocyanates and related amines IV. 2,4- and 2,6-toluene diamine in hydrolyzed plasma and urine after test-chamber exposure of humans to 2,4- and 2,6-toluene diisocyanate. Int. Arch. Occup. Environ. Health 63:253-259.

Budavari, S., M.J. O'Neil, A. Smith, P.E. Heckelman, and J.F. Kinneary, eds. 1996. The Merck Index, 12th Ed. Whitehouse Station, NJ: Merck and Co., Inc. Pp. 1626.

Bugler, J., R.L. Clark, I.D. Hill, and M. McDermott. 1991. The acute and long-term respiratory effects of aromatic di-isocyanates. A five year longitudinal study of polyurethane foam workers. III Report No. 10848. International Isocyanate Institute.

Butcher, B.T., R.M. Karr, C.E. O'Neil, M.R. Wilson, V. Dharmarajan, J.E. Salvaggio, and H. Weill. 1979. Inhalation challenge and pharmacologic studies of toluene diisocyanate (TDI)-sensitive workers. J. Allergy Clin. Immunol. 64:146-152.

Butcher, B.T., C.E. O'Neil, M.A. Reed, J.E. Salvaggio, and H. Weill. 1982. Development and loss of toluene diisocyanate reactivity: Immunologic, pharmacologic, and provocative challenge studies. J. Allergy Clin. Immunol. 70:231-235.

CMA (Chemical Manufacturers Association). 1989. Comments of the Diisocyanates Program Panel of the Chemical Manufacturers Association Regarding the Potential Carcinogenicity of Toluene Diisocyanate. Comments for NIOSH Current Intelligence Bulletin Toluene Diisocyanate. Washington, DC: CMA.

CPS (Chemical Products Synopsis). 2001. Toluene Diisocyanate. Adams, NY: Mannsville Chemical Products Corp. April 2001.

Diem, J.E., R.N. Jones, D.J. Hendrick, et al. 1982. Five-year longitudinal study of workers employed in a new toluene diisocyanate manufacturing plant. Am. Rev. Resp. Dis. 126:420-428.

Di Stefano, A., M. Saetta, P. Maestrelli, G. Milani, F. Pivirotto, C.E. Mapp, and L.M. Fabbri. 1993. Mast cells in the airway mucosa and rapid development of occupational asthma induced by toluene diisocyanate. Am. Rev. Respir. Dis. 147:1005-1009.

Duncan, B., L.D. Scheel, E.J. Fairchild, R. Killens, and S. Graham. 1962. Toluene diisocyanate inhalation toxicity: Pathology and mortality. Am. Ind. Hyg. Assoc. J. 23:447-456.

Elkins, H.B., G.W. McCarl, H.G. Brugsch, and J.P. Fahy. 1962. Massachusetts experience with toluene di-isocyanate. Am. Ind. Hyg. Assoc. J. 23:265-272.

EPA (U.S. Environmental Protection Agency). 1995. 1993 Toxics Release Inventory. Office of Pollution Prevention and Toxics, U.S. Environmental Protection Agency, Washington, D.C.

EPA (U.S. Environmental Protection Agency). 1996. 2,4-/2,6-Toluene diisocyanante mixture. Integrated Risk Information System (IRIS) [Online]. Available: http://www.epa.gov/iris/subst/0503.htm [October 1, 1996].

Fabbri, L.M., P. Boschitto, E. Zocca, G. Milani, F. Pivirotto, M. Plebani, A. Burlina, B. Licata, and C.E. Mapp. 1987. Bronchoalveolar neutrophilia during late asthmatic reactions induced by toluene diisocyanate. Am. Rev. Resp. Dis. 136:36-42.

Gagnaire, F., B. Ban, J.C. Micillino, M. Lemonnier, and P. Bonnet. 1996. Bronchial responsiveness and inflammation in guinea-pigs exposed to toluene diisocyanate: A study on single and repeated exposure. Toxicology 114:91-100.

Garabrant, D.H., and S.P. Levine. 1994. A critical review of the methods of exposure assessment and the pulmonary effects of TDI and MDI in epidemiologic studies. Final report to the Diisocyanates Panel, Chemical Manufacturers Association, Washington, DC.

German Research Association (Deutsche Forschungsgemeinschaft). 2000. List of MAK and BAT Values, 2000. Commission for the Investigation of Health Hazards of Chemical Compounds in the Work Area, Report No. 36. Weinheim, Federal Republic of Germany: Wiley VCH.

Hama, G.M. 1957. Symptoms in workers exposed to isocyanates. AMA Arch. Indust. Health 16:232-233.

Hartung, R. 1994. Cyanides and nitriles. Pp. 3163-3165 in Patty's Industrial Hygiene and Toxicology, 4th Ed., G.D. Clayton and F.E. Clayton, eds. New York, NY: John Wiley & Sons, Inc.

Henschler, D., W. Assman, and K.-O. Meyer. 1962. On the toxicology of toluene diisocyanate [in German]. Archiv für Toxikologie 19:364-387.

Horspool, G.M., and J.E. Doe. 1977. Toluene Di-isocyanate: Acute Inhalation Toxicity in the Rat. Study No. HR0082, Report No. CTL/T/1097.Imperial Chemicals Industries Limited, Central Toxicology Laboratory, Cheshire, UK.

Hosein, H.R., and S. Farkas. 1981. Risk associated with the spray application of polyurethane foam. Am. Ind. Hyg. Assoc. J. 42:663-665.

Huang, J., X. Wang, B. Chen, X. Zhou, and T. Matsushita. 1993. Dose-response relationships for chemical sensitization from TDI and DNCB. Bull. Environ. Contam. Toxicol. 51:732-739.

IARC (International Agency for Research on Cancer). 1985. Some Chemicals Used in Plastics and Elastomers. IARC Monographs on the Evaluation of the Carcinogenic Risk of Chemicals to Humans, Vol. 39. Lyon, France: IARC. Pp. 287-323.

Karol, M.H., C. Dixon, M. Brady, and Y. Alarie. 1980. Immunologic sensitization and pulmonary hypersensitivity by repeated inhalation of aromatic isocyanates. Toxicol. Appl. Pharmacol. 53:260-270.

Karol, M.H. 1983. Concentration-dependent immunologic response to toluene diisocyanate (TDI) following inhalation exposure. Toxicol. Appl. Pharmacol. 68:229-241.

Karol, M.H. 1986. Respiratory effects of inhaled isocyanates. CRC Crit. Rev. Toxicol. 16:349-379.

Kennedy, A.L., M.F. Stock, Y. Alarie, and W.E. Brown. 1989. Uptake and distribution of ^{14}C during and following inhalation exposure to radioactive toluene diisocyanate. Toxicol. Appl. Pharmacol. 100:280-292.

Kimmerle, G. 1976. Acute inhalation toxicity of diisocyanates, polymer isocyanates and coating systems on rats. Report No. 6200. Institute for Toxicology, Wuppertal-Elberfeld, Germany. Pp. 32.

Kociba, R.J., D.G. Keyes, and E.L. Wolfe. 1979. Histopathological observations on selected tissues of Syrian hamsters exposed by inhalation to vapors of toluene diisocyanate (TDI) for 6 hours/day, 5 days/week for 4 weeks. Report No. NA-A-12. The Dow Chemical Company, Midland, MI. Pp. 5.

Linaweaver, P.G. 1972. Prevention of accidents resulting from exposure to high concentrations of foaming chemicals. J. Occup. Med. 14:24-30.

Loeser, E. 1983. Long-term toxicity and carcinogenicity studies with 2,4/2,6-toluene-diisocyanate (80/20) in rats and mice. Toxicol. Lett. 15:71-81.

Maestrelli, P., A. Di Stefano, P. Occari, G. Turato, G. Milani, F. Pivirotto, C.E. Mapp, L.M. Fabbri, and M. Saetta. 1995. Cytokines in the airway mucosa of subjects with asthma induced by toluene diisocyanate. Am J. Respir. Crit. Care Med. 151:607-612.

Maxon, F.C. 1964. Respiratory irritation from toluene diisocyanate. Arch. Environ. Health 8:755-758.

Ministry of Social Affairs and Employment (SDU Uitgevers). 2000. Nationale MAC (Maximum Allowable Concentration) List, 2000. Ministry of Social Affairs and Employment, The Hague, The Netherlands.

Moller, D.R., R.T. McKay, I.L. Bernstein, and S.M. Brooks. 1986. Persistent airways disease caused by toluene diisocyanate. Am. Rev. Respir. Dis. 134:175-6.

NCI (National Cancer Institute). 1979. Bioassay of 2,4-diaminotoluene for possible carcinogenicity. Technical Report Series No. 162, CAS No. 95-80-7, NCI-CR-TR-162. National Cancer Institute, National Institutes of Health, Bethesda, MD.

NIOSH (National Institute for Occupational Safety and Health). 1996. Toluene 2,4-Diisocyanate. Documentation for Immediately Dangerous to Life or Health

Concentrations (IDLHs). NIOSH, Cincinnati, OH [Online]. Available: http://www.cdc.gov/niosh/idlh/584849.html.

NIOSH (National Institute for Occupational Safety and Health). 1997. NIOSH Pocket Guide to Chemical Hazards. NIOSH, Cincinnati, OH [Online]. Available: http://www.cdc.gov/niosh/npg/ npgd0621.html.

NRC (National Research Council). 1993. Guidelines for Developing Community Emergency Exposure Levels for Hazardous Substances. Washington, DC: National Academy Press.

NRC (National Research Council). 2001. Standing Operating Procedures for Developing Acute Exposure Guideline Levels for Hazardous Chemicals. Washington, DC: National Academy Press.

NTP (National Toxicology Program). 1986. Toxicology and carcinogenesis studies of commercial grade 2,4 (80%)- and 2,6 (20%)-toluene diisocyanate (CAS No. 26471-62-5) in F344/N rats and $B6C3F_1$ mice (gavage studies). NTP-TR-251. National Toxicology Program, National Institute of Environmental Health Sciences, Research Triangle Park, NC.

Owen, P.E. 1983. The toxicity and carcinogenicity to rats of toluene diisocyanate vapour administered by inhalation for a period of 113 weeks Addendum report volume 1. Report No. 2507-484/1. Hazleton Laboratories Europe Ltd., Harrogate, England.

Sangha, G.K., and Y. Alarie. 1979. Sensory irritation by toluene diisocyanate in single and repeated exposures. Toxicol. Appl. Pharmacol. 50:533-547.

Scheerens, H., T.L. Buckley, T. Muis, H. Van Loveren, and F.P. Nijkamp. 1996. The involvement of sensory neuropeptides in toluene diisocyanate-induced tracheal hyperreactivity in the mouse airways. Br. J. Pharmacol. 119:1665-1671.

Shiotsuka, R.N. 1987a. Sensory irritation study of MONDUR TDS in male Sprague-Dawley rats. Study no. 86-341-01. Stilwell, KS: Mobay Corporation.

Shiotsuka, R.N. 1987b. Sensory irritation study of MONDUR TD-80 in Sprague-Dawley rats. Study no. 86-341-02. Stilwell, KS: Mobay Corporation.

Skarping, G., T. Brorson, and S. Carsten. 1991. Biological monitoring of isocyanates and related amines III. Test chamber exposure of humans to toluene diisocyanate. Int. Arch. Occup. Environ. Health 63:83-88.

Stevens, M.A., and R. Palmer. 1970. The effect of tolylene diisocyanate on certain laboratory animals. Proc. R. Soc. Med. 63:380-382.

ten Berge, W.F., A. Zwart, and L.M. Appelman. 1986. Concentration-time mortality response relationship of irritant and systemically acting vapours and gases. J. Hazard. Mat. 13:301-309.

Timchalk, C., F.A. Smith, and M.J. Bartels. 1992. Metabolic fate of ^{14}C-toluene 2,4-diisocyanate in Fischer 344 rats. File number 10910. International Isocyanate Institute. Dow Chemical Co., Midland, MI.

Timchalk, C., F.A. Smith, and M.J. Bartels. 1994. Route-dependent comparative metabolism of [^{14}C]toluene 2,4-diisocyanate and [^{14}C]toluene 2,4-diamine in Fischer 344 rats. Toxicol. Appl. Pharmacol. 124:181-190.

Tyl, R.W. 1988. Developmental toxicity study of inhaled toluene diisocyanate vapor in CD (Sprague-Dawley) rats. Project NA-AB-50. International Isocyanate Institute. Bushy Run Research Center, Export, PA.

Tyl, R.W., and T.L. Neeper-Bradley. 1989. Two-generation reproduction study of inhaled toluene diisocyanate in CD (Sprague-Dawley) rats. Final report (51-576, dated 17 March 1989). Project NA-AB-50. International Isocyanate Institute. Bushy Run Research Center, Export, PA.

Vandenplas, O., A. Cartier, H. Ghezzo, Y. Cloutier, and J.-L. Malo. 1993. Response to isocyanates: Effect of concentration, duration of exposure, and dose. Am. Rev. Respir. Dis. 147:1287-1290.

Warren, D.L. 1994a. Respiratory sensitization study in guinea pigs: Controls challenged with 2,4- or 2,6- toluene diisocyanate (TDI). Project AM-AB-89-II. File number 11120. International Isocyanate Institute. Miles, Inc., Stilwell, KS.

Warren, D.L. 1994b. Respiratory sensitization study with 2,4- or 2,6-toluene diisocyanate (TDI) in guinea pigs. Project AM-AB-89. File number 10992. International Isocyanate Institute. Miles, Inc., Stilwell, KS.

Wazeter, F.X. 1964a. Six-hour acute inhalation toxicity study in rats. No. 100-012. International Research and Developmental Corp.

Wazeter, F.X. 1964b. Acute inhalation exposure in male albino rats. No. 203-006. International Research and Developmental Corp.

Weyel, D.A., B.S. Rodney, and Y. Alarie. 1982. Sensory irritation, pulmonary irritation, and acute lethality of a polymeric isocyanate and sensory irritation of 2,6-toluene diisocyanate. Toxicol. Appl. Pharmacol. 64:423-430.

WHO (World Health Organization). 1987. Toluene diisocyanates. Environmental Health Criteria 75. Geneva: World Health Organization.

Wilson, R.H., and G.L. Wilson. 1959. Toxicology of diisocyanates. J. Occup. Med. 1:448-450.

Wong, K.-L., M.H. Karol, and Y. Alarie. 1985. Use of repeated CO_2 challenges to evaluate the pulmonary performance of guinea pigs exposed to toluene diisocyanate. J. Toxicol. Environ. Health 15:137-148.

Woolrich, P.F. 1982. Toxicology, industrial hygiene and medical control of TDI, MDI and PMPPI. Am. Ind. Hyg. Assoc. J. 43:89-97.

Zapp, J.A. 1957. Hazards of isocyanates in polyurethane foam plastic production. Arch. Ind. Health 15:324-330.

APPENDIX A

LC_{50} Probit Plot

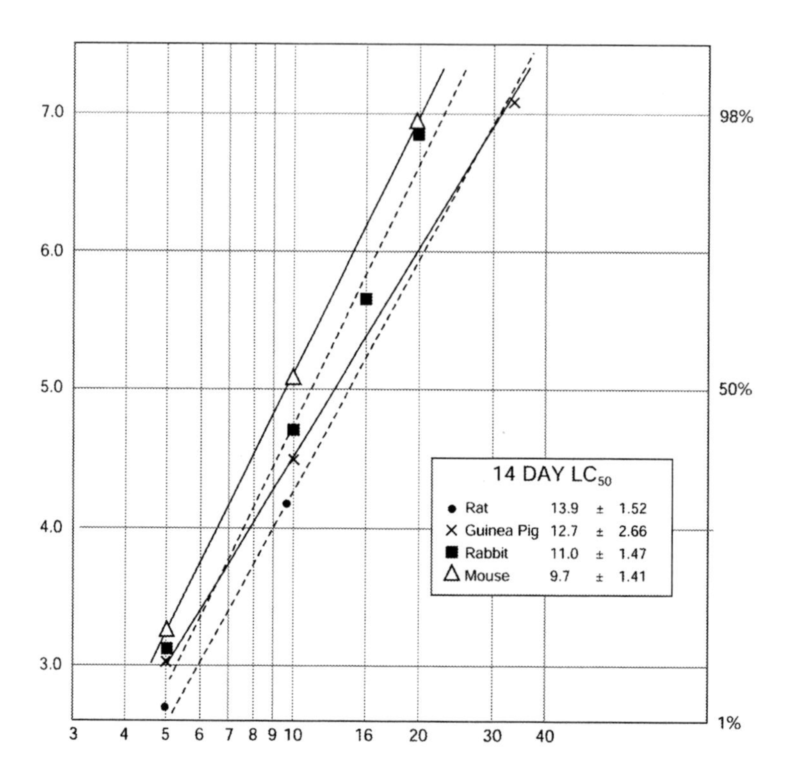

FIGURE 4A-1 LC_{50} probit plot. Source: Duncan et al. 1962. Reprinted with permission from the *American Industrial Hygiene Association Journal*; copyright 1962, AIHA.

APPENDIX B

Derivation of AEGL Values

Derivation of AEGL-1

Key study: Baur 1985

Toxicity end point: Asthmatic subjects experienced cough, rhinitis, chest tightness, dyspnea, throat irritation, and/or headache from exposure at 0.01 ppm for 1 h, and then, after a rest, at 0.02 ppm for another hour.

Time-scaling: None

Uncertainty factors: None—asthmatic people are considered a sensitive population

Modifying factor: None

10-min AEGL-1: 0.02 ppm

30-min AEGL-1: 0.02 ppm

1-h AEGL-1: 0.02 ppm

4-h AEGL-1: 0.01 ppm

8-h AEGL-1: 0.01 ppm

Derivation of AEGL-2

Key study: Henschler et al. 1962

Toxicity end points: Severe eye and throat irritation in humans exposed at 0.5 ppm for 30 min

Time-scaling: $C^n \times t = k$ (ten Berge et al. 1986; NRC 2001), $n = 3$ for extrapolating to the 10-min time point, $n = 1$ for extrapolating to the 1-, 4-, and 8-h time points

Uncertainty factors: 3 for intraspecies variability (not protecting hypersusceptible individuals)

Calculations: *10-min time point*
$(C/\text{UFs})^3 \times t = k$
$(0.5/3)^3 \times 0.5\ \text{h} = 0.0023\ \text{ppm}^3\cdot\text{h}$

1-, 4-, and 8-h time points
$(C/\text{UFs})^1 \times t = k$
$(0.5/3)^1 \times 0.5\ \text{h} = 0.083\ \text{ppm}\cdot\text{h}$

10-min AEGL-2: $(0.0023\ \text{ppm}^3\cdot\text{h}/0.167\ \text{h}) = 0.24$ ppm

30-min AEGL-2: 0.5 ppm/3 = 0.17 ppm

1-h AEGL-2: $(0.083\ \text{ppm}\cdot\text{h}/1\ \text{h}) = 0.083$ ppm

4-h AEGL-2: $(0.083\ \text{ppm}\cdot\text{h}/4\ \text{h}) = 0.021$ ppm

8-h AEGL-2: 0.021 ppm

Derivation of AEGL-3

Key Study: Duncan et al. 1962

Toxicity end point: The 4-h LC_{50} of 9.7 ppm in mice was used for derivation of AEGL-3 values. An approximate threshold for lethality is obtained by dividing the LC_{50} by 3.

Time-scaling: $C^n \times t = k$ (ten Berge et al. 1986)

	$n = 3$ for extrapolating to the 10-min, 30-min, and 1-h time points; $(3.23)^3 \times 4.0 = 135.21$ ppm·h
	$n = 1$ for extrapolating to the 8-h time point; $(3.23)^1 \times 4.0 = 12.92$ ppm·h
Uncertainty factors:	10 (3 for intraspecies variability and 3 for interspecies variability
Calculations:	*10-min, 30-min, and 1-h time points* $(C/\text{UFs})^3 \times t = k$ $(3.23/10)^3 \times 4\text{ h} = 0.135\text{ ppm}^3\cdot\text{h}$
	8-h time point $(C/\text{uncertainty factors})^1 \times t = k$ $(3.23/10)^1 \times 4\text{ h} = 1.292$ ppm·hr
10-min AEGL-2:	0.65 ppm
30-min AEGL-2:	$(0.135\text{ ppm}^3\cdot\text{h}/0.5\text{ h}) = 0.65$ ppm
1-h AEGL-2:	$(0.135\text{ ppm}^3\cdot\text{h}/1\text{ h}) = 0.51$ ppm
4-h AEGL-2:	(3.23 ppm/10) = 0.32 ppm
8-h AEGL-2:	(1.292 ppm·h/8 h) = 0.16 ppm

APPENDIX C

DERIVATION SUMMARY

ACUTE EXPOSURE GUIDELINE LEVELS FOR TOLUENE 2,4- AND 2,6-DIISOCYANATE (CAS Nos. 584-84-9 and 91-08-7)

AEGL-1				
10 min	30 min	1 h	4 h	8 h
0.02 ppm	0.02 ppm	0.02 ppm	0.01 ppm	0.01 ppm
Key reference: Baur, X. 1985. Isocyanate hypersensitivity. Final report to the International Isocyanate Institute. III File No. 10349; III Project: E-AB-19.				
Test species/strain/number: Human subjects, gender not given;10 healthy controls and 15 asthmatics				
Exposure route/concentrations/durations: Inhalation; 0.02 ppm for 2 h (controls); 0.01 ppm for 1 h, 45 min rest, 0.02 ppm for 1 h (asthmatics)				
Effects: Controls—significant increase in airway resistence (R_{aw}) immediately and 30 min after beginning of exposure; eye irritation and/or cough. Asthmatics—no change in lung function parameters; chest tightness, rhinitis, cough, dyspnea, throat irritation, and/or headache				
End point/concentration/rationale: Some (5/15) asthmatic humans exposed for 1 h at 0.01 ppm and, after a 45 min rest, at 0.02 ppm for another hour experienced chest tightness, rhinitis, cough, dyspnea, throat irritation, and/or headache.				
Uncertainty factors/rationale: Total uncertainty factor: None Interspecies: Not applicable, human data used Intraspecies: 1, asthmatics were used as the test population				
Modifying factor: None				
Animal to human dosimetric adjustment: Not applicable				
Time-scaling: Extrapolation to time points was not conducted. Because the asthmatics tolerated 0.02 ppm for 1 h after pre-exposure at 0.01 ppm, it is assumed that this population could tolerate the lower concentration for a longer duration.				
Data quality and support for the AEGL values: AEGL-1 values are considered conservative and should be protective of the toxic effects of TDI outside those expected as defined under AEGL-1.				

AEGL-2				
10 min	30 min	1 h	4 h	8 h
0.24 ppm	0.17 ppm	0.083 ppm	0.021 ppm	0.02 ppm
Key reference: Henschler, D., Assman, W., and Meyer, K.-O. 1962. On the toxicology of toluenediisocyanate [in German]. Archiv. für Toxikologie 19:364-387				
Test species/strain/number: Human, healthy male, 6				
Exposure route/concentrations/durations: 0.01-1.3 ppm 2,4/2,6-, 2,4-, or 2,6-TDI for 30 min				
Effects: Effects were similar for both isomers and the mixture. 0.1 ppm: eye and nose irritation; ≥0.5 ppm: marked discomfort, lacrimation, nasal secretion (determinant for AEGL-2); 1.3 ppm: intolerable.				
End point/concentration/rationale: Humans exposed at 0.5 ppm for 30 min experienced pronounced irritation (marked discomfort, lacrimation, nasal secretion)				
Uncertainty factors/rationale: Total uncertainty factor: 3 Interspecies: Not applicaple, human data used Intraspecies: 3. The use of a higher uncertainty factor would make the AEGL-2 values similar to AEGL-1 values, which are based on levels that asthmatic humans can tolerate.				
Modifying factor: Not applicable				
Animal to human dosimetric adjustment: Not applicable				
Time-scaling: $C^n \times t = k$ where n ranges from 0.8 to 3.5 (ten Berge et al. 1986). In the absence of an emphirically derived, chemical-specific exponent, scaling was performed using $n = 1$ for extrapolating to the 10-min time point and $n = 3$ for the 1- and 4-h time points. The 4-h value is also used for the 8-h value because extrapolation to 8 h resulted in a concentration similar to that causing mild effects in polyurethane foam sprayers exposed for >7 h (Hosein and Farkas 1981) and in manufacturing workers on 8-h shifts (Hama 1957).				
Data quality and support for the AEGL values: Some individuals with pre-existing bronchial hyper-reactivity have been shown to respond to TDI with nonpathologic bronchial obstruction (4/15), but no significant differences were observed in lung function parameters. AEGL-2 values also supported by animal data.				

AEGL-3				
10 min	30 min	1 h	4 h	8 h
0.65 ppm	0.65 ppm	0.51 ppm	0.32 ppm	0.16 ppm
Key reference: Duncan, B., Scheel, L.D., Fairchild, E.J., Killens, R., and Graham, S. 1962. Toluene diisocyanate inhalation toxicity: pathology and mortality. Am. Indus. Hygiene Assoc. J. 23:447-456.				
Test species/strain/number: Mice, 120 total animals				
Exposure route/concentrations/durations: Inhalation, 0.1, 1.0, 2, 5, 10, 20, or 34 ppm for 4 h				
Effects: 9.7 ppm 4-h LC_{50} in the mouse: concentration dependent signs of toxicity included mouth breathing, lacrimation, salivation, and restlessness; Histopathologic examination of surviving animals: coagulation necrosis and desquamation of the superficial epithelial lining of the trachea and major bronchi, cleared by day 7 post-exposure in the 2 ppm group.				
End point/concentration/rationale: 3.23 ppm is an estimated lethality threshold obtained by dividing the 4-h mouse LC_{50} by 3. That is approximately equal to the LC_{01} obtained by extrapolating the probit plot in the Duncan et al. (1962) paper.				
Uncertainty factors/rationale: Total uncertainty factor: 10. Interspecies: 3. The LC_{50} was determined in the rat, guinea pig, rabbit, and mouse. The 4-h LC_{50} values ranged from 9.7 ppm in the mouse to 13.9 ppm in the rat. These results argue for low variability between species. In addition, the use of a higher uncertainty factor would place the AEGL-3 levels in the range of the AEGL-2 values, which were set based on human data. The most sensitive species, the mouse, was used to derive the AEGL-3 values. Intraspecies: 3. Use of a greater uncertainty factor would result in values below those supported by human data for AEGL-3 effects.				
Modifying factor: Not applicable				
Animal to human dosimetric adjustment: Not applicable				
Time-scaling: $C^n \times t = k$ where n ranges from 0.8 to 3.5 (ten Berge et al. 1986). In the absence of an empirically derived, chemical-specific exponent, scaling was performed using $n = 1$ for extrapolating to the 30-min and 1-h time points and $n = 3$ for the 8-h time point. The 10-min AEGL-3 value was flatlined from the 30-min value.				
Data quality and support for the AEGL values: Presensitized individuals might exist in the general population, but the rate of sensitization cannot be predicted. If the rate of sensitization to TDI in the general population were				

AEGL-3 *Continued*
quantifiable, the committee might have considered lower values for AEGL-3. At the AEGL-3 levels, individuals who have a stronger reaction to TDI might not be protected from severe effects. The mouse appears to be the most sensitive species tested, although LC_{50} values did not vary greatly.

5

Uranium Hexafluoride[1]

Acute Exposure Guideline Levels

SUMMARY

Uranium hexafluoride (UF_6) is a volatile solid. It is one of the most highly soluble industrial uranium compounds and, when airborne, hydrolyzes immediately on contact with water to form hydrofluoric acid (HF) and uranyl fluoride (UO_2F_2) as follows:

$$UF_6 + 2H_2O \rightarrow UO_2F_2 + 4HF.$$

Thus, an inhalation exposure to UF_6 is actually an inhalation exposure to a mixture of fluorides. The HF component may produce pulmonary irritation, corrosion, or edema, and the uranium component may produce renal injury.

---

[1]This document was prepared by the AEGL Development Team comprising Cheryl Bast (Oak Ridge National Laboratory) and National Advisory Committee (NAC) on Acute Exposure Guideline Levels for Hazardous Substances member George Rusch (Chemical Manager). The NAC reviewed and revised the document and AEGL values as deemed necessary. Both the document and the AEGLs were then reviewed by the National Research Council (NRC) Subcommittee on Acute Exposure Guideline Levels. The NRC subcommittee concluded that the AEGLs developed in this document are scientifically valid conclusions on the basis of the data reviewed by the NRC and are consistent with the NRC guidelines reports (NRC 1993, 2001).

If death does not occur as a result of HF exposure, renal effects from the uranium moiety might occur (Just 1984).

In the absence of relevant chemical-specific data for deriving the lowest acute exposure guideline level (AEGL) values for UF_6, modifications of the AEGL-1 values for HF were used. The use of HF as a surrogate for UF_6 was deemed appropriate for the development of AEGL-1 values, which are based on irritation symptoms, because it is likely that HF, a hydrolysis product, is responsible for those low-level effects. The HF AEGL-1 values were based on the threshold for pulmonary inflammation in healthy human adults (Lund et al. 1999). Because a maximum of 4 moles (mol) of HF are produced for every mole of UF_6 hydrolyzed, a stoichiometric adjustment factor of 4 was applied to the HF AEGL-1 values to approximate AEGL-1 values for UF_6; the AEGL-1 values for UF_6 are constant across time up to 1 hour (h) because the HF AEGL-1 values were held constant across time. AEGL-1 values for UF_6 were derived for only the 10-minute (min), 30-min, and 1-h time points because derivation of 4- and 8-h values resulted in AEGL-1 values greater than the 4- and 8-h AEGL-2 values calculated for UF_6. That would be inconsistent with the total database.

The AEGL-2 values were based on renal pathology in dogs exposed to UF_6 at 192 milligrams per cubic meter (mg/m^3) for 30 min (Morrow et al. 1982). An uncertainty factor (UF) of 3 was used to extrapolate from animals to humans, and a UF of 3 was also applied to account for sensitive individuals (total UF = 10). This total UF of 10 is considered sufficient because the use of a larger total UF would yield AEGL-2 values below or approaching the AEGL-1 values, which are considered no-observed-effect levels and were based on a threshold for inflammation in humans. Furthermore, humans were exposed to HF repeatedly at up to 8 parts per million (ppm) with only slight nasal irritation; that is stoichiometrically equivalent to a UF_6 exposure at 28.8 mg/m^3, a concentration equivalent to the 10-min AEGL-2. The concentration-exposure time relationship for many irritant and systemically acting vapors and gases may be described by $C^n \times t = k$ (C = concentration, t = time, and k is a constant), where the exponent n ranges from 0.8 to 3.5 (ten Berge et al. 1986). To obtain protective AEGL values in the absence of an empirically derived, chemical-specific scaling exponent, temporal scaling was performed using $n = 3$ when extrapolating to shorter time points and $n = 1$ when extrapolating to longer time points. (Although a chemical-specific exponent of 0.66 was derived from a rat lethality study in which the end point was pulmonary edema, the default values were used for time-scaling AEGL-2 values, because the end points for AEGL-2 [renal toxicity] and for death [pulmonary edema] involve different mechanisms of action.)

The AEGL-3 was based on an estimated 1-h lethality threshold in rats (one-third the LC_{50} [concentration lethal to 50% of subjects] of 365 mg/m^3) (Leach et al 1984). That approach is considered appropriate due to the steepness ($n = 0.66$) of the concentration-response curve for lethality in rats exposed to UF_6. An intraspecies UF of 3 was applied and is considered sufficient because the steep concentration-response curve for lethality implies little intra-individual variability. A UF of 3 was also applied for interspecies variability (total UF = 10). An application of the full interspecies UF of 10 would result in AEGL-3 values that are inconsistent with the overall data set. For example, assuming the production of 4 mol of HF from the hydrolysis of 1 mol of UF_6 (1 mg/m^3 = 0.0695 ppm), exposure at the proposed AEGL-3 values would be equivalent to exposure to HF at 60 ppm for 10-min, 20 ppm for 30-min, 10 ppm for 1-h, 2.5 ppm for 4-h, and 1.3 ppm for 8-h. No effects on respiratory parameters were noted in healthy humans after exposure to HF at 6.4 ppm for 1-h or after repeated exposures of up to 8.1 ppm (Lund et al. 1999; Largent 1960, 1961). Although this scenario does not account for the uranium portion of the exposure, given the steepness of the concentration-response curve it is unlikely that people would experience life-threatening effects at these concentrations. Therefore, an interspecies UF of 3 is justified. The value was then scaled to the 10- and 30-min and 4- and 8-h time points using $C^1 \times t = k$. An exponent of 0.66 was derived from rat lethality data collected from exposures ranging from 2 min to 1 h duration in the key study. The exponent was rounded to 1.0 for extrapolation because the data used to derive the exponent were limited (from one study) and the derived n was below the range of a normal dose-response curve.

The calculated AEGLs are listed in Table 5-1.

1. INTRODUCTION

UF_6 is a volatile solid. It may present both chemical and radiological hazards. It is one of the most highly soluble industrial uranium compounds and, when airborne, hydrolyzes rapidly on contact with water to form hydrofluoric acid (HF) and uranyl fluoride (UO_2F_2) as follows:

$$UF_6 + 2H_2O \rightarrow UO_2F_2 + 4HF.$$

Thus, an inhalation exposure to UF_6 is actually an inhalation exposure to a mixture of fluorides. Chemical toxicity may involve pulmonary irritation, corrosion, or edema from the HF component and/or renal injury from the

TABLE 5-1 Summary of Proposed AEGLs for Uranium Hexafluoride (mg/m^3)

Classification	10 min	30 min	1 h	4 h	8 h	End Point (Reference)
AEGL-1 (Nondisabling)	3.6	3.6	3.6	NR	NR	Modification of hydrogen fluoride AEGL-1 values (EPA 2001)
AEGL-2 (Disabling)	28	19	9.6	2.4	1.2	Renal tubular pathology in dogs (Morrow et al. 1982)
AEGL-3 (Lethal)	216	72	36	9.0	4.5	Estimated 1-h NOEL for death in the rat (Leach et al. 1984)

Abbreviations: mg/m^3, milligrams per cubic meter; NOEL, no-observed-effect level; NR, not recommended.

uranium component (Fisher et al. 1991). There is also evidence to suggest that the fluoride ion might contribute to renal toxicity after repeated exposure (G. Rusch, Honeywell Inc., personal commun., Nov. 24, 1999). As concentration decreases and duration increases, the effects of HF are reduced, and the effects of the uranium component may increase (Spiegel 1949). Thus, if death does not occur from HF exposure, renal effects from the uranium moiety may occur (Just 1984).

In addition, UF_6 emits alpha, beta, and gamma radiation (Allied Signal 1994). The predominant uranium isotope found in nature, ^{238}U, is not fissionable, and consequently, for uranium to be used as a fuel in nuclear reactors, the ratio of ^{235}U to ^{238}U is increased from 0.72% to 2-4% (enriched) by the gaseous diffusion process. UF_6 is produced by fluorinating processed uranium oxide ore (U_3O_8), and is used in the uranium enrichment process. The UF_6 stream containing both ^{235}U and ^{238}U is divided into separate streams. One is enriched, and the other is depleted (ATSDR 1997).

The predominant concern from acute exposure to soluble uranium compounds is chemical toxicity. However, depending on the degree of enrichment, radiologic toxicity might also be important. However, it is difficult to quantitate the radiological hazard without knowing the precise degree of enrichment (ATSDR 1997).

The physicochemical properties of UF_6 are presented in Table 5-2.

TABLE 5-2 Physical and Chemical Data for Uranium Hexafluoride

Parameter	Value	Reference
Common name	Uranium hexafluoride	ATSDR 1997
Synonyms	Uranium fluoride (fissile); UN2977	ATSDR 1997
CAS registry no.	7783-81-5	ATSDR 1997
Chemical formula	UF_6	ATSDR 1997
Molecular weight	352.02	ATSDR 1997
Physical state	Colorless solid	ATSDR 1997
Odor	Odorless	AIHA 1995
Vapor pressure	115 mm Hg at 25°C	ATSDR 1997
Density	4.68 g/cm^2 at 21°C	ATSDR 1997
Triple (melting) point	64.05°C at 760 mmHg	AIHA 1995
Boiling point	56.2°C	ATSDR 1997
Solubility	Soluble in carbon tetrachloride and chloroform; hydrolyzes in cold water, alcohol, and ether	AIHA 1995
Conversion factors in air	1 ppm = 14.4 mg/m^3 (in dry air) 1 mg/m^3 = 0.0695 ppm (in dry air) 1 mg of UF_6 = 0.676 mg of U 1 mg of UO_2F_2 = 0.767 mg of U (based on a mixture of ^{235}U and ^{238}U)	AIHA 1995

2. HUMAN TOXICITY DATA

2.1. Acute Lethality

2.1.1. Case Reports

On September 2, 1944, an accidental UF_6 release occurred at a Manhattan Engineering District pilot plant located at the Philadelphia naval shipyard (Kathren and Moore 1986). A weld ruptured in a 2.4 m × 0.2 m cylinder containing gaseous UF_6, causing the cylinder to act as a rocket. It traveled 164 m, tore out pipes and fittings in its path, and released an estimated 182 kg of UF_6. The released UF_6 and steam produced a dense cloud of UF_6

and hydrolysis products that enveloped the area within 91.4 m of the rupture site. The cloud rapidly dispersed and exposure duration was estimated to be 17 seconds. Twenty workers were exposed at varying degrees, depending on their location during the rupture. Two people, both in a direct line with the cylinder, died—one died 10-16 min after exposure, and the other died 70 min after exposure. Three people were seriously injured, 12 were hospitalized for observation, and three were without symptoms. The seriously injured individuals experienced chemical conjunctivitis with edema, chemical erosion of the cornea (resulting in temporary blindness), first-, second-, and third-degree chemical burns, nausea and vomiting, chemical bronchitis, pulmonary edema, and/or shock. Blood counts suggested hemoconcentration in one seriously injured individual, and albumin and casts were observed in the urine. Another seriously injured individual had an essentially normal blood profile; however, albuminuria and urine casts were noted. The seriously injured workers completely recovered within 3 weeks (wk) of the accident. Two of the seriously injured workers were re-examined 38 years (y) after the accident for deposition and potential long-term effects of uranium exposure. The initial lung deposition was estimated to be 40-50 mg of uranium on the basis of fragmentary urinary excretion data obtained shortly after the accident, and the initial long-term bone deposition was estimated at 410 µg (5.2 Bq) in the worker exposed at the highest concentration. That would result in a 40-y dose-equivalent to the bone of approximately 200 mrem. The 38-y follow-up medical and health physics exams of two of the seriously injured workers revealed no detectable uranium deposition and no findings attributable to uranium exposure (Kathren and Moore 1986; Moore and Kathren 1985).

Another accident occurred on January 4, 1986, at a uranium conversion facility in Gore, Oklahoma (Nuclear Regulatory Commission 1986). A 12,700 kg overloaded cylinder containing approximately 13,500 kg of UF_6 ruptured when it was being heated in a steam chest to remove excess UF_6. The release lasted for approximately 40 min. A dense white cloud formed and was pushed by the wind. It quickly engulfed the process building (the facility's main structure) and formed a plume expanding to the south-southeast. When the UF_6 left the ruptured cylinder, not all of it reacted immediately to form UO_2F_2 and HF; pieces of solid UF_6 were scattered widely around the steam chest area. Any UF_6 remaining when the area was sprayed down reacted with the water and was likely captured in the water spray. It was estimated that approximately half of the released UF_6 was washed into an emergency pond. Therefore, approximately 6,700 kg contributed to the noxious white cloud. When completely reacted with water,

6,700 kg of UF_6 forms 5,900 kg of UO_2F_2 and 1,500 kg of HF. Peak 10-min UO_2F_2 concentrations ranged from 0.011 mg/m^3 to 8.8 mg/m^3, and peak 10-min HF concentrations ranged from 0.008 mg/m^3 to 2.4 mg/m^3. One-hour average uranium concentrations ranged from <0.052 µg/m^3 to 20,000 µg/m^3. It was assumed that 4.55 × 106 g of uranium was released into the atmosphere in a 45 min period. The isotopic composition was 1.49 Ci of ^{234}U, 0.07 Ci of ^{235}U, and 1.49 Ci of ^{238}U.

There were 42 workers at the plant site when the accident occurred. Seven of the workers (contract workers) were in a trailer well away from and upwind of the release point. One worker (an operator in a scrubber building 50 ft from the steam chest) died from pulmonary edema produced by HF inhalation within a few hours of the exposure. Another was treated for skin irritation and burns from HF exposure, and 21 others were examined at a hospital and kept overnight for observation. Four of the 21 workers were released the following morning, 14 were kept more than 1 day (d) and were given oral sodium bicarbonate to enhance uranium excretion, and three were transferred to another hospital for observation and treatment of potential lung damage from HF exposure. Of the eight employees who returned to spray water on the fumes, one was in the group transferred to another hospital with serious respiratory symptoms, and five were hospitalized for more than 1 d (including the individual with HF burns). Only two of those workers were released on the day of the accident.

Local off-site residents were also asked to report to a hospital for examination. One resident was kept overnight for observation, but no medical treatment was deemed necessary. Urine samples were collected from all on-site workers and from 100 off-site members of the public for uranium bioassay and urinalysis (osmolality, creatinine, protein, glucose, LDH, albumin, β-microglobulin, and *n*-acetyl-β-glucosaminidase). Initial uranium concentrations in the urine of employees (noncontract workers) ranged from <2 µg/L to 11,000 µg/L; however, those values might not be indicative of the total amount of uranium excreted, because the total sample volumes were less than 1 L. Levels in the urine among the contracted employees ranged from <2 µg/L to 2,600 µg/L. The initial mean for all workers was 2,165 µg/L. The initial mean for those that returned to the scene of the release to wash down the area was 2,898 µg/L, and the initial mean for the contract workers in the trailer was 456 µg/L. These data were used to calculate the following estimated uranium intakes. The noncontracted employees had mean estimated intakes of 0.26 mg to 27.63 mg, and contracted employees had intakes ranging from 0.02 mg to 7.73 mg. The average for all workers was 6.5 mg. Urinalysis values showed elevated glucose levels in five workers, increased β-microglobulin levels in nine other workers, and increased *n*-acetyl-β-glucosaminidase in seven workers (three of whom also had increased β-microglobulin and one of whom showed increased glucose). However, there was no correlation between estimated uranium intake and urinalysis values, and those data are further confounded by the fact that several workers were diagnosed with urinary tract infections and/or diabetes (both incidental to the accident) during their medical examinations. Therefore, no conclusions can be drawn concerning possible renal damage from the UF_6 exposure.

The uranium bioassay data for the 100 off-site individuals suggest estimated intakes of 0.1 mg to 0.9 mg of uranium. Effective total-body dose equivalents and doses to organs were estimated assuming a 1-h exposure to the plume. The hypothetically maximally exposed off-site individual would have received a total-body dose of 6.5 millirem (mrem), 43.3 mrem to the kidneys, 202.5 mrem to the bone, and 2.5 mrem to the lungs. The nearest residents would have received a total-body dose of 2.2 mrem, 14.6 mrem to the kidneys, 68.2 mrem to the bone, and 0.8 mrem to the lungs. The maximally exposed worker would have received 46 mrem to the total body and 1,400 mrem to the bone. The background radiation total-body dose for persons living in the Gore, Oklahoma, area is 106 mrem/y. Thus, for maximally exposed off-site persons, residents, and even for exposed workers, the doses are too small to produce any acute health effects from radiation. The likelihood of long-term (carcinogenic) effects is discussed in Section 2.5.

2.2. Nonlethal Toxicity

2.2.1. Case Reports

Both lethal and nonlethal and effects from UF_6 exposure were identified in two accident reports (Nuclear Regulatory Commission 1986; Kathren and Moore 1986). Those case reports are described in Section 2.1.1.

2.3. Developmental and Reproductive Toxicity

No information concerning developmental and reproductive toxicity in humans following acute inhalation exposure to UF_6 was located.

2.4. Genotoxicity

Genotoxicity studies regarding acute human exposure to UF_6 were not available.

2.5. Carcinogenicity

No information concerning carcinogenicity in humans following acute inhalation exposure to UF_6 was located. Because UF_6 is radioactive, it could potentially damage DNA and lead to cancer; however, without knowing the precise degree of enrichment, it is difficult to quantitate the potential radiologic hazard. Lung cancers observed in uranium mine workers are believed to be the result of chronic radon exposure (ATSDR 1997).

A 38-y follow-up examination of two workers seriously injured from an accidental acute UF_6 exposure revealed no findings (carcinogenic or noncarcinogenic) attributable to uranium exposure (Moore and Kathren 1985).

The estimated lifetime cancer risk from the Gore, Oklahoma, accident was calculated for maximally exposed off-site individuals and nearby residents assuming a 1-h exposure to the plume and additional exposure by inhalation of resuspended material (annual inhalation) and ingestion of potentially contaminated vegetables, meat, and milk (annual ingestion) (Nuclear Regulatory Commission 1986). The assessment took into account the specific activities of the uranium isotopes. The calculated cancer fatality risks were extremely small—4.3×10^{-7} to the maximally exposed individual and 1.4×10^{-7} to a typical nearby resident.

2.6. Summary

Case reports from human accidental exposures to UF_6 indicate that acute toxicity is chemical, not radiologic, in nature and is due to the hydrolysis products, hydrogen fluoride (HF) and uranyl fluoride (UO_2F_2). At high concentrations, death from HF-induced pulmonary edema is observed. Severe ocular injury; skin burns; and ocular, mucous membrane, and respiratory irritation are also attributable to HF. Kidney damage attributable to UO_2F_2, was also suggested from urinalysis data. No developmental and reproductive or genotoxicity data were available. The carcinogenic hazard from radiation exposure is negligible compared with the chemical toxicity from acute inhalation exposure to UF_6.

3. ANIMAL TOXICITY DATA

3.1. Acute Lethality

3.1.1. Rats

Groups of 20 rats (strain and gender not specified) were exposed to UF_6 at 942, 1,083, or 2,284 mg/m^3 for 10 min and observed for up to 30 d (Spiegel 1949). Animals were exposed in a 4-ft^3 chamber made of Monel metal that was set 3 ft above the floor. Small portable cages were placed in two layers within the chamber. UF_6 was introduced into the chamber as a continuous stream. Six 1-min samples were taken during the 10 min period and analyzed for concentration using a ferrocyanide acetic acid chemical method. Upon removal from the chamber, all animals were gasping for breath; severe irritation of the nasal passages and conjunctivitis was also observed and persisted for a period of several hours. No rats died during the exposure; mortality during the 30 d after exposure was 10%, 30%, and 75%, respectively, for the 942-, 1,083-, or 2,284-mg/m^3 exposure groups (Table 5-3). The maximum incidence of mortality was observed from 5 d to 8 d post-exposure, with few animals dying after day 10. Additional rats sacrificed in a daily serial study after the termination of exposure exhibited severe damage to the renal cortical tubules. This renal damage peaked between 5 and 8 d after exposure, and then regenerative processes started. Mild, inflammatory pulmonary changes were observed in rats sacrificed 1-4 d after exposure. One rat that died on day 5 exhibited severe lung injury.

In 46 single exposures, Leach et al. (1984) exposed groups of 10 male Long-Evans rats (200-250 g) to concentrations of UF_6 ranging from 651 mg/m^3 to 409,275 mg/m^3 for 2, 5, or 10 min. Two forms of UF_6 were used in the experiments; the first was prepared from natural uranium and the second was enriched to 94% ^{235}U. Exposures were nose-only in a polyethylene and stainless steel chamber. The UF_6 was metered from a heated gas cylinder into the inlet air system of the exposure unit. Before introduction into the exposure chamber, the concentrated vapor was mixed with clean compressed air maintained at a relative humidity of 50-95%. The UF_6 was drawn into the chamber and past the individual test animals' noses. The chamber airflow was maintained at 16-28 L/min at a temperature of 22-

TABLE 5-3 Mortality After 10-Min Uranium Hexafluoride Exposure (%)

	Rat		Mouse		Guinea Pig	
	Mortality		Mortality		Mortality	
Concentration (mg/m^3)	During Exposure	During 30 d After	During Exposure	During 30 d After	During Exposure	During 30 d After
942	0	10	0	20	0	13
1,083	0	30	0	70	0	20
2,284	0	75	30	95	27	40

Source: Data from Spiegel et al. 1949.

26°C. During exposure, a colorimetric, cascade impactor method was used to monitor uranium concentration, and fluoride ion was measured with a fluoride ion-specific electrode. Immediately after exposure, the animals were removed from the restraining tubes and their heads were washed with aqueous detergent to remove deposited uranium and HF. Selected animals were whole-body counted for radioactivity. Surviving animals were returned to individual metabolism cages and held for up to 14 d. Urine and feces were collected daily, and when enriched uranium was used, excreta was measured for uranium content by gamma counting. The assessment of renal injury was monitored (only in exposures with 0-20% mortality) by measuring daily excretory changes in urine volume, protein, glucose, and selected enzymes (NAG, GGT, AP, LDH, AST, ALT).

Mortality data are presented in Tables 5-4, 5-5, and 5-6. Of the rats that died, 25% died during exposure or within 48 h of exposure, 59% died between days 3 and 7, and 17% died between days 8 and 14. The study authors attributed deaths occurring within 48 h of exposure to HF toxicity. Deaths occurring after 48 h were attributed to renal (uranium) toxicity or a combination of uranium and HF toxicity. Calculated rat LC_{50} values were 117,515 mg/m^3, 57,100 mg/m^3, and 17,751 mg/m^3 for the 2-, 5-, and 10-min time points, respectively. The mean percentages of absorbed uranium dose found in urine, bone, kidneys, and lungs were 13-61%, 28-79%, 3-9%, and 0-6%, respectively. Cumulative uranium excretion over 6 d in the feces ranged from 59% to 81% of the inhaled dose. With a soluble uranium compound such as UF_6, one might have expected predominately urinary excretion (especially given that fecal excretion typically accounts for less than 1% of excretion of soluble uranium compounds after inhalation expo-

TABLE 5-4 Rat Mortality and Renal Injury Indicators After 2-Min Uranium Hexafluoride Exposure

UF_6 Concentration (mg/m^3)	Mortality at 14-d Post-Exposure	Renal Injury Indicators of Surviving Animals
651	0/10	P, G, PO
3,151	0/10	P, G, PO
11,925[a]	1/10	P, G, PO
14,216[a]	0/10	P, G, PO
17,899	0/10	P, G, PO
24,689	0/10	P, G, PO, AP, LDH
44,615	0/10	P, G, PO
60,533	1/10	P, G, PO
98,979	2/10	P, G, PO
116,050	3/10	P, G, PO
140,533	7/10	—
153,107	6/10	—
172,411	2/10	P, G, PO
251,479	8/10	—
352,663	9/10	—
409,275	9/10	—

[a]Enriched UF_6 (94% ^{235}U), other exposures were natural uranium.
Abbreviations: AP, alkaline phosphatase; G, glucosuria; LDH, lactate dehydrogenase; P, proteinuria ; PO, polyuria.
Source: Data from Leach et al. 1984.

sure); however, it is likely that the rats trapped the particles in the nasopharyngeal region and swallowed them. Urine bioassays indicated renal injury at all exposure durations (Tables 5-4, 5-5, and 5-6). Proteinuria and glucosuria were greatest 4-5 d following exposure and tended to disappear during the second week. Urine volumes reached their maximum on days 5 and 6 post-exposure. Histopathologic examination of kidneys (14-d post-exposure) showed renal lesions at all concentrations except for the two lowest 2-min exposures. Regeneration of tubular epithelium was apparent, and many tubules were dilated and lined with flattened abnormal epithelium. The lungs of surviving rats examined 14-d post-exposure showed no

TABLE 5-5 Rat Mortality and Renal Injury Indicators After 5-Min Exposure

UF_6 Concentration (mg/m^3)	Mortality at 14-d Post-Exposure	Renal Injury Indicators of Surviving Animals
859	0/10	P, G, PO, UP
3,669	1/10	P, G, PO
4,675[a]	0/10	—
9,260[a]	0/10	PO
9,571[a]	1/10	P, G, PO
20,000	1/10	P, G, PO
25,178[a]	0/10	—
25,562	1/10	PO
26,686	6/10	—
26,938[a]	4/10	—
30,429	8/10	—
37,899	8/10	—
39,260[a]	1/10	P, G, PO
41,731[a]	6/10	—
54,940	1/10	PO
65,237	5/10	—
81,731	8/10	—

[a]Enriched UF_6 (94% ^{235}U), other exposures were natural uranium.
Abbreviations: G, glucosuria; P, proteinuria; PO, polyuria; UP, five urinary enzymes (GGT, LDH, AP, LAP, NAG).
Source: Data from Leach et al. 1984.

treatment-related effects. However, animals that died during or shortly after exposure had congestion, acute inflammation, and focal degeneration of the upper respiratory tract. The tracheas, bronchi, and lungs exhibited acute inflammation with epithelial degeneration, acute bronchial inflammation, and acute pulmonary edema and inflammation, respectively.

Leach et al. (1984) also reported a 1-h rat LC_{50} of 1,095 mg/m^3. No further information waspresented, although it is assumed that exposure conditions were similar to those used in the 2-min, 5-min, and 10-min studies.

TABLE 5-6 Rat Mortality and Renal Injury Indicators After 10-Min Exposure

UF_6 Concentration (mg/m^3)	Mortality at 14-d Post-Exposure	Renal Injury Indicators of Surviving Animals
932[a]	0/10	P, G, PO, NAG
5,947	0/10	P, G, PO
8,388[a]	1/10	P, G, PO, NAG
10,059	2/10	P, G, PO
10,074	1/10	P, G, PO
16,006	3/10	P, G, PO
16,435	1/10	P, G, PO
17,766	8/10	—
21,464	6/10	PO
22,855	10/10	—
31,302	10/10	—
36,568	8/10	—
57,899	7/10	PO
89,334	10/10	—

[a]Enriched UF_6 (94% ^{235}U), other exposures were natural uranium.
Abbreviations: G, glucosuria; NAG, *n*-acetyl glucosamine; P, proteinuria; PO, polyuria.
Source: Data from Leach et al. 1984.

3.1.2. Mice

Groups of 20 mice (strain and gender not specified) were exposed to UF_6 at 942, 1,083, or 2,284 mg/m^3 for 10 min and observed for up to 30 d (Spiegel 1949). Animals were exposed, in 4-ft^3 chamber made of Monel metal that was set 3 ft above the floor. Small portable cages were placed in two layers within the chamber. The UF_6 was introduced into the chamber as a continuous stream. Six 1-min samples were taken during the 10-min period and were analyzed for concentration using a ferrocyanide acetic acid chemical method. Upon removal from the chamber, all animals were gasping for breath; severe irritation of the nasal passages, conjunctivitis, and closed, encrusted eyes were also observed and persisted in survivors for a period of several hours. Six mice exposed at the highest concentration died during the exposure period; mortality during the 30 d after exposure was 20%, 70%, and 95%, respectively, for the 942-, 1,083-, or 2,284-mg/m^3 exposure groups (Table 5-3). The maximum incidence of mortality was observed from day 5 through day 8 post-exposure, with few animals dying after day 10.

3.1.3. Guinea Pigs

Groups of 15 guinea pigs (strain and gender not specified) were exposed to UF_6 at 942, 1,083, or 2,284 mg/m^3 for 10 min and observed for up to 30 d (Spiegel 1949). Animals were exposed in a 4-ft^3 chamber made of Monel metal that was set 3 ft above the floor. Small portable cages were placed in two layers within the chamber. The UF_6 was introduced into the chamber as a continuous stream. Six 1-min samples were taken during the 10-min period and analyzed for concentration using a ferrocyanide acetic acid chemical method. Upon removal from the chamber, all animals were gasping for breath; severe irritation of the nasal passages and conjunctivitis were also observed and persisted in survivors for a period of several hours. Four guinea pigs in the high-concentration group died during the exposure period; mortality during the 30 d after exposure was 13%, 20%, and 40%, respectively, for the 942-, 1,083-, and 2,284-mg/m^3 exposure groups (Table 5-3). The maximum incidence of mortality was observed from day 5 through day 8 post-exposure, and few animals died after day 10.

In 14 single exposures, Leach et al. (1984) exposed groups of six male Hartley guinea pigs (approximately 350 g) to concentrations of UF_6 ranging from 26,642 mg/m^3 to 203,580 mg/m^3 for 2 min. Two forms of UF_6 were used in the experiments; the first was prepared from natural uranium and the second was enriched to 94% ^{235}U. Exposures were nose-only in a polyethylene and stainless steel chamber. The UF_6 was metered from a heated gas cylinder into the inlet air system of the exposure unit. Before introduction into the exposure chamber, the concentrated vapor was mixed with clean compressed air maintained at a relative humidity of 50-95%. The UF_6 was drawn into the chamber and past the individual test animals noses. The chamber airflow was maintained at 16-28 L/min at a temperature of 22-26°C. During exposure, a colorimetric cascade impactor method was used to monitor uranium concentration, and fluoride ion was measured with a fluoride ion-specific electrode. Immediately after exposure, the animals were removed from the restraining tubes and their heads were washed with aqueous detergent to remove deposited uranium and HF. Selected animals were whole-body counted for radioactivity. Surviving animals were returned to individual metabolism cages and held for up to 14 d. Urine and feces were collected daily, and when enriched uranium was used, excreta were measured for uranium content by gamma counting. The assessment of renal injury was monitored (only in exposures with 0-20% mortality) by measuring daily excretory changes in urine volume, protein, glucose, and selected enzymes (NAG, GGT, AP, LDH, AST, ALT). Mortality data are presented in Table 5-7. Of the guinea pigs that died, 64% died during exposure or within 48 h of exposure; 31% died between days 3 and 7; and 6% died between days 8 and 14. Deaths occurring within 48 h of exposure were attributed to HF toxicity. Deaths occurring after 48 h were attributed to renal (uranium) toxicity or a combination of uranium and HF toxicity. The calculated guinea pig 2-min LC_{50} value was 91,864 mg/m^3 (Table 5-8). The mean percent of absorbed uranium dose found in urine, bone, kidneys, and lungs was 56-59%, 34-43%, 2-5%, and 0-5%, respectively. Cumulative uranium excretion over 6 d in the feces was approximately 59% of the inhaled dose. With a soluble uranium compound such as UF_6, one might have expected predominately urinary excretion; however, it is likely that the guinea pigs trapped the particles in the nasopharyngeal region and swallowed them. Urine bioassays indicated renal injury at all exposure concentrations (Table 5-7).

3.2. Nonlethal Toxicity

3.2.1. Dogs

Morrow et al. (1982) exposed two young adult female beagle dogs (6-10 kg) to UF_6 at 192-284 mg/m^3 for 30 min to 1 h. The dogs had been anesthetized with intravenous pentobarbital sodium and placed into individual body plethysmographs with their heads exteriorized through a neck seal. Volume meters were connected to the plethysmograph and the dogs were intubated with a cuffed endotracheal tube. The chambers used in the recent studies of Leach et al. (1984) were also used for this dog study; however, in place of the rat/guinea pig holders being connected to the chamber, the dog's endotracheal tube was connected using a stopper. One dog was necropsied 6 d post-exposure, and the other 16 d post-exposure. No lung pathology was observed. Renal pathology showed tubular regeneration typified by increased mitotic figures and development of flattened tubular epithelium. Residual manifestations of tubular dilation were observed in the dog sacrificed 16 d after exposure.

TABLE 5-7 Guinea Pig Mortality and Renal Injury Indicators After 2-Min Exposure

UF_6 Concentration (mg/m^3)	Mortality at 14-d Post-Exposure	Renal Injury Indicators of Surviving Animals
26,642	0/6	P, G, PO
34,083[a]	2/6	PO
39,896	1/6	P, G, PO, NAG
40,459	1/6	PO
60,429[a]	4/6	—
61,317	3/6	P, G, PO
73,092	3/6	PO
96,376	1/6	PO
98,979	3/6	P, G, PO
112,944	3/6	PO
140,533	5/6	—
153,107	3/6	P, G, PO
156,464	4/6	—
203,580	6/6	—

[a]Enriched UF_6 (94% ^{235}U), other exposures were natural U.
Sources: G, glucosuria; NAG, *n*-acetyl glucosamine; P, proteinuria; PO, polyuria.
Source: Data from Leach et al. 1984.

In this same study, 17 dogs were similarly exposed to UO_2F_2 at 200-270 mg/m^3 for 30 min to 2.5 h and sacrificed from 1 d to 19 d post-exposure. No lung damage was noted in these dogs. Between 1 d and 3 d, kidneys

TABLE 5-8 Calculated LC_{50} Values for Animals Exposed to Uranium Hexafluoride

Species	Exposure Time (min)	LC_{50} (mg/m^3)
Rat	2	177,515
Rat	5	57,100
Rat	10	17,751
Rat	60	1,095
Guinea Pig	2	91,864

Source: Data from Leach et al. 1984.

showed widely scattered necrosis of segments of the deep convoluted tubules and the straight portions of the corticomedullary junctions extending toward the mid-cortex. The tubules contained hyaline material and proteinaceous casts, and some collecting tubules contained calcified debris. The tubular epithelial cells were pale and were often anucleated and denuded. After 3 d, evidence of tubular regeneration was rare, but after 6 d, it was seen in all dogs, especially at lower concentrations. In animals sacrificed 14 d and 19 d after exposure, tubular dilation and atrophy were present.

3.3. Developmental and Reproductive Toxicity

No studies were located concerning developmental and reproductive toxicity from acute inhalation of UF_6 in animals.

3.4. Genotoxicity

No genotoxicity studies were located concerning UF_6 acute inhalation exposures.

3.5. Carcinogenicity

Inhalation of UF_6 at 0.05 mg/m^3 for 1 y caused mild renal tubular effects in one of 13 dogs and one of 141 rats. Mild atrophic changes were observed in "most dogs and in some rats and rabbits." No neoplastic effects were noted (Spiegel 1949), and no control data were presented.

3.6. Summary

Rats and guinea pigs exposed to UF_6 for 2-10 min showed upper and lower respiratory tract irritation, pulmonary edema, and renal lesions. Those effects are the results of the hydrolysis products of UF. HF is responsible for the respiratory effects, and uranium is responsible for the renal pathology. Dogs administered UF_6 intratracheally showed no respiratory or pulmonary effects but showed renal effects from uranium. Lethality data suggest that rats are more resistant to UF_6-induced lethality than are guinea pigs (Table 5-8). No reproductive and developmental data or genotoxicity data were located. There was no evidence of carcinogenicity in dogs that inhaled 0.05 mg/m^3 for 1 y.

4. SPECIAL CONSIDERATIONS

4.1. Metabolism and Disposition

UF_6 and its hydrolysis products, hydrogen fluoride (HF) and uranyl fluoride (UO_2F_2), are all soluble and may be absorbed from the lung after inhalation exposure.

UO_2F_2 is rapidly absorbed from the lung into the bloodstream as the UO_2^{2+} ion behaves like calcium and complexes with serum proteins and bicarbonate. The uranyl ion forms biscarbonate, $UO_2(CO_3)_3^-$, and citrate complexes in plasma. The UO_2^{2+} also binds to red blood cells. As the UO_2^{2+} ion passes through the kidney and is filtered from blood, it dissociates within the tubular filtrate and recombines with cell surface ligands. Removal of the uranyl ion from the kidneys is described by a two-component exponential model where 92-95% is excreted in the urine with a half-time of 2-6 d and the remainder is excreted with a half-time of 30-340 d. While most of the uranium absorbed into the blood is excreted in urine, there is some generalized systemic deposition in soft tissues, primarily the liver, and in the bone, where it replaces calcium. Deposition rates in bone correlate with bone growth and remodeling rates. Recirculating UO_2^{2+} from bone goes to the kidneys. Long-term retention of uranium in bone is described by a two-compartment model, where 90% of uranium is cleared in a half-time of 20-300 d and the remainder is cleared in a half-time of 3,700-5,000 d (Fisher et al. 1991).

HF, the other UF_6 hydrolysis product, is highly soluble in water and is readily absorbed in the upper respiratory tract at lower concentrations. However, at very high concentrations HF may reach the lungs. The relatively low dissociation constant (3.5×10^{-4}) allows the non-ionized compound to penetrate the skin, respiratory system, or gastrointestinal system and form a reservoir of fluoride ions that bind calcium and magnesium, forming insoluble salts (Bertolini 1992). Fluoride ion is readily absorbed into the blood stream and is carried to all organs of the body in proportion to their vascularity and the concentration in the blood; equilibrium across biological membranes is rapid (Perry et al. 1994). Significant deposition occurs in the bone, where the fluoride ion substitutes for the hydroxyl group of hydroxyapatite, the principal mineral component of bone. In humans, chronic exposure to elevated levels of fluoride or HF has produced osteofluorosis. Elimination is through the kidneys.

4.2. Mechanism of Toxicity

The mechanism of toxicity of UF_6 involves both chemical and radiologic components. The chemical toxicity is due to the two hydrolysis products, UO_2F_2 and HF. Uranium metal has five oxidation states; however, only the +4 and +6 are stable enough to be of practical importance. The +6 forms the uranyl ion, which forms the water soluble compounds and is an important species of uranium in body fluids (ATSDR 1999).

Nephrotoxicity is the most sensitive indicator of toxicity from acute high-level exposure to uranium in mammals. The kidney effects are manifested histologically as glomerular and tubular wall degeneration, and ultrastructural examination indicates damage to endothelial cells in the glomerulus, such as loss of cell processes and decreased density of the endothelial fenestrae. Loss of the brush border, cellular vacolization, and necrosis may be noted in the terminal segments of the proximal convoluted tubules. This process may lead to the disruption of tubular resorption of solutes and to a decreased filtration rate of the glomerulus. Indicators of uranium-induced nephrotoxicity include excessive urinary excretion of protein, glucose, enzymes, catalase, or alkaline phosphatase.

Bicarbonate activity in the kidneys has been suggested as a possible mechanism for uranium-induced renal toxicity. Uranium is usually combined with bicarbonate or a plasma protein in the blood. When uranium is released from the bicarbonate in the kidneys it is free to form complexes with phosphate ligands and protein in the tubular wall, which may result in damage. However, the uranium is not tightly bound and is thus released within days. Within a week post-exposure, the uranium is cleared from the kidneys, and the tubules start to regenerate. Another proposed mechanism of uranium-induced renal toxicity involves inhibition of sodium transport-dependent and transport-independent ATP utilization and mitochondrial oxidative phosphorylation in the proximal tubule (ATSDR 1999).

There is also evidence to suggest that the fluoride ion might contribute to renal toxicity, especially after repeated exposures. Renal cortex degeneration and necrosis was noted in rats exposed to fluoride (as HF) at 24 mg/m^3 for 6 h/d, 6 d/wk for 30 d; however, no renal effects were observed in rats similarly exposed at 7 mg/m^3 (Stokinger 1949). Renal tubular necrosis and congestion were reported in rats and guinea pigs exposed to HF (unreported concentration and duration) (Machle et al. 1934). Degeneration and necrosis was noted in the convoluted tubules of rabbits exposed to fluoride (as HF) at 14 mg/m^3 for 6-7 h/d for 50 d (Machle and Kitzmiller 1935). The relevance of HF inhalation exposure to kidney toxicity in humans is questionable, because the occupational exposure data suggest that HF is not nephrotixic in humans (ATSDR 1993).

The available studies indicate that HF is a severe irritant to the skin, eyes, and respiratory tract. It is particularly irritating to the anterior nasal passages where, depending on species and concentration, it appears to be effectively scrubbed from the inhaled air. Effective deposition in the anterior nasal passages may be attributed to the high solubility and reactivity of HF. Penetration into the lungs results in pulmonary hemorrhage and edema and may result in death. Although renal and hepatic changes have been observed in animal studies, serious systemic effects are unlikely to occur from an acute exposure.

UF_6 emits alpha, beta, and gamma radiation and, thus, may interact with and damage DNA and chromosomes. If not repaired, that damage may eventually lead to cancer. The amount of radiation emitted is dependent on the degree of enrichment of the UF_6. In the case of acute inhalation exposures to the soluble UF_6, short-term radiotoxicity would be negligible compared with chemical toxicity, even in the case of highly enriched UF_6 (Just and Elmer 1984). The United Nations Scientific Committee on the Effects of Atomic Radiation (UNSCEAR 1993) has concluded that naturally occurring and depleted uranium, although radioactive, present primarily a toxicologic, rather than radiologic hazard, because the hypothetical radiologic toxicity predicted in skeletal tissues has not been observed in humans or experimental animals.

4.3. Susceptible Populations

People with impaired renal function (chemical toxicity of uranium), fetuses and developing neonates (radiologic hazard), and asthmatic individuals (HF toxicity) might be especially susceptible to UF_6 toxicity.

4.4. Species Variability

Acute inhalation studies suggest that, with regard to UF_6-induced lethality, mice are most sensitive, rats are moderately sensitive, and guinea pigs are most resistant (Spiegel 1949; Leach et al. 1984). In a subchronic study,

TABLE 5-9 Mortality for Animals Exposed to Uranium Hexafluoride[a] (%)

Concentration (mg/m^3)	Guinea Pigs	Dogs	Rats	Mice	Rabbits
0.3	0	0	5	5	0
3	5	20	0	92	80
20	45	40	75	100	100

[a]Animals were exposed for 6 h/d for 30 d.
Source: Data from Spiegel 1949.

Spiegel (1949) exposed adult mice, rats, guinea pigs, rabbits, and dogs to UF_6 at 0.3-20 mg/m^3 for 6 h/d for 30 d (Table 5-9, above). With regard to lethality in adult animals, guinea pigs and dogs were least susceptible, rats were moderately susceptible, and mice and rabbits were highly susceptible. However, those relative species sensitivities must be viewed with caution, because firm concentration-response relationships were not established (5% of rats died at 0.3 mg/m^3, but 0% died at 3 mg/m^3). With regard to body weight (Table 5-10), the greatest weight loss in the subchronic study was observed in rabbits, followed by guinea pigs; dogs and rats showed similar weight loss patterns.

4.5. Temporal Extrapolation

The concentration-exposure time relationship for many irritant and systemically acting vapors and gases has been described by $C^n \times t = k$, where the exponent, n, typically ranges from 0.8 to 3.5 (ten Berge et al. 1986). When the rat 2-60 min LC_{50} values from Leach et al. (1984) are considered in the derivation of the exponent, a value of 0.66 is obtained (Figure 5-1).

4.6. Other Information

In a previous attempt to assess the toxicity of UF_6 and its hydrolysis products to humans, four toxicologists were asked to use the available animal data to calculate levels for human health effects (Just and Elmer 1984; Just 1984). The four individuals independently derived values using four different approaches. The group then convened and reached a set of con-

TABLE 5-10 Percent Weight Change in Animals Exposed to Uranium Hexafluoride[a]

Concentration (mg/m^3)	Guinea Pigs	Dogs	Rats	Rabbits
0.3	+11	+9	+3	0
3	+1	-12	-8	-12
20	-13	-3	-6	-21
0.075[b]	+11	-2	+70	—
0.3[b]	—	+10	+75	+3

[a]Animals were exposed for 6 h/d for 30 d.
[b]Very young, growing animals.

sensus values for 1-h exposures. Assumptions used in the derivation were as follows: a 70-kg man, a resting respiration rate of 7.5 L/min, and an absorption factor of 0.43. The ICRP (International Commission on Radiological Protection) method was used to model absorbed uranium to the inhalation exposure concentration. The derived estimates are shown in Table 5-11.

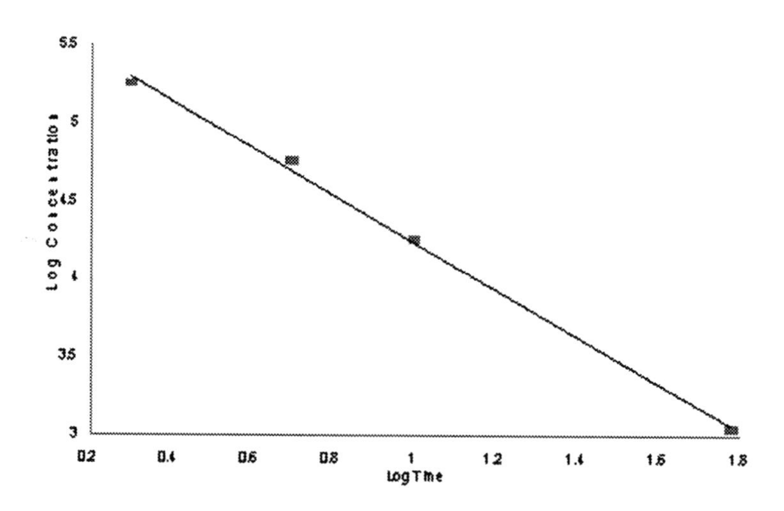

FIGURE 5-1 Best fit concentration-time curve.

TABLE 5-11 Human Toxicity Estimates for a 1-h Inhalation Exposure to Uranium Hexafluoride

Effect	Concentration (mg/m^3)
No effect	<9.6
Mild health effects	9.6-18.5
Renal injury	18.5
10% lethal	352
50% lethal	862

Sources: data from Just and Elmer 1984; Just 1984.

5. RATIONALE AND PROPOSED AEGL-1

5.1. Human Data Relevant to AEGL-1

No human exposure data consistent with the definition of AEGL-1 were available for UF_6. However, the work of Just and Elmer (1984), which estimates a 1-h no-observed-effect level (NOEL) of <9.6 mg/m^3, supports the derivation of the AEGL-1 values presented below.

5.2. Animal Data Relevant to AEGL-1

Effects observed from inhalation exposure to UF_6 in experimental animals were more severe than those defined by AEGL-1.

5.3. Derivation of AEGL-1

In the absence of relevant chemical-specific data for the derivation of AEGL-1 values, modifications of the AEGL-1 values for HF were used to derive the AEGL-1 values for UF_6. The use of HF as a surrogate for UF_6 was deemed appropriate for the development of AEGL-1 values because it is likely that HF, a hydrolysis product of UF_6, is responsible for the low level irritation effects of UF_6 relevant to the AEGL-1 definition. The HF AEGL-1 values were based on the threshold for pulmonary inflammation in healthy human adults (Lund et al. 1999). Because a maximum of 4 mol of HF are produced for every mole of UF_6 hydrolyzed, a stoichiometric adjustment factor of 4 was applied to the HF AEGL-1 values to approximate AEGL-1 values for UF_6; the AEGL-1 values for UF_6 are constant across time up to 1 h because the HF AEGL-1 values were held constant across time. AEGL-1 values were derived for only the 10-min, 30-min, and 1-h time points, because derivation of the 4- and 8-h AEGL-1s resulted in values greater than the 4- and 8-h AEGL-2s for UF_6, which would be inconsistent with the total database. The AEGL-1 values for UF_6 are presented in Table 5-12, and the calculations for those AEGL-1 values are presented in Appendix A. The relationship between AEGL-1 and animal exposures is shown in Figure 5-2.

The values are supported by the estimated 1-h human NOEL of <9.6 mg/m^3 (Just and Elmer 1984; Just 1984). Applying an uncertainty factor (UF) of 3 for sensitive individuals, a value of 3.2 mg/m^3 is obtained, which is in close agreement with the values derived for AEGL-1.

6. RATIONALE AND PROPOSED AEGL-2

6.1. Human Data Relevant to AEGL-2

Case reports describing accidental human exposures to UF_6 that led to effects consistent with the definition of AEGL-2 exist. However, because of the lack of reliable concentration and duration parameters, those data are not appropriate for derivation of AEGL-2 values.

6.2. Animal Data Relevant to AEGL-2

Two dogs exposed to UF_6 at 192-284 mg/m^3 for 30 min to 1 h exhibited renal tubular pathology 6 d and 16 d post-exposure (Morrow et al. 1982).

6.3. Derivation of AEGL-2

The renal pathology observed in dogs was used as the basis for AEGL-2 values. The lowest concentration (192 mg/m^3) and the shortest exposure

TABLE 5-12 AEGL-1 Values for Uranium Hexafluoride (mg/m^3)

10 min	30 min	1 h	4 h	8 h
3.6	3.6	3.6	NR	NR

Abbreviation: NR, not recommended.

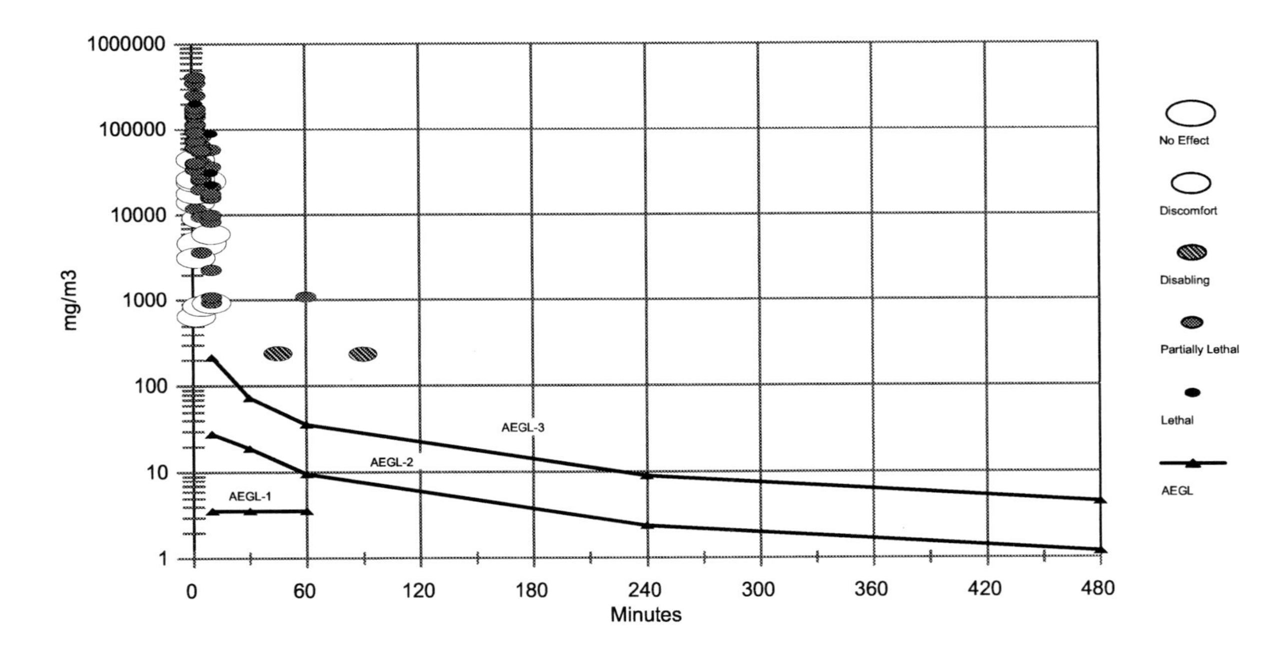

FIGURE 5-2 Category plot for AEGL values and effects of UF_6 on animals.

TABLE 5-13 AEGL-2 Values for Uranium Hexafluoride (mg/m^3)

10 min	30 min	1 h	4 h	8 h
28	19	9.6	2.4	1.2

time (30 min) were used in the derivation. Although only one dog was exposed at each treatment level, the use of this data is supported by the Morrow et al. (1982) study in which 17 dogs exposed to UO_2F_2 at 200-270 mg/m^3 for 30 min to 2.5 h exhibited similar renal pathology. Although the UO_2F_2 is not the title compound, it is the hydrolysis product of UF_6 likely responsible for the renal effects. A UF of 3 was used to extrapolate from animals to humans, and a UF of 3 was also applied to account for sensitive individuals (total UF = 10). This total UF is considered sufficient because the use of a larger total UF would yield AEGL-2 values below or approaching the AEGL-1 values, which for HF were considered to be no-observed-effect levels and were based on a threshold for inflammation in humans (HF study). Also, humans were exposed repeatedly to HF at up to 8 ppm with only slight nasal irritation; that concentration is stoichiometrically equivalent to UF at 28.8 mg/m^3. The concentration-exposure time relationship for many irritant and systemically acting vapors and gases may be described by $C^n \times t = k$, where the exponent, n, ranges from 0.8 to 3.5 (ten Berge et al. 1986). To obtain protective AEGL values in the absence of an empirically derived, chemical-specific scaling exponent, temporal scaling was performed using $n = 3$ when extrapolating to shorter time points and $n = 1$ when extrapolating to longer time points. (Although a chemical-specific exponent of 0.66 was derived from rat lethality data, the default values were utilized for time-scaling AEGL-2 values because the end points for AEGL-2 [renal toxicity] and for death [pulmonary edema] involve different mechanisms of action.) The AEGL-2 values for UF_6 are presented in Table 5-13, above, and the calculations for those AEGL-2 values are presented in Appendix A. The relationship between AEGL-2 and animal exposures is shown in Figure 5-2.

7. RATIONALE AND PROPOSED AEGL-3

7.1. Human Data Relevant to AEGL-3

Deaths from accidental exposure to UF_6 were reported in two case reports. However, because of the lack of reliable concentration and duration parameters, those data are not appropriate for derivation of AEGL-3 values.

7.2. Animal Data Relevant to AEGL-3

Well-conducted LC_{50} studies are available in rats, mice, and guinea pigs (Spiegel 1949; Leach et al. 1984). However, the majority of those studies used exposure durations of 2-, 5-, or 10-min and are, thus, inappropriate (too short) for extrapolating to the longer AEGL time periods. The exception is a 1-h rat LC_{50} study (Leach et al. 1984).

7.3. Derivation of AEGL-3

An estimated 1-h lethality threshold (one-third the LC_{50}) in the rat is 365 mg/m^3. That value was used as the basis for deriving the AEGL-3 values. This approach is considered appropriate due to the steepness ($n = 0.66$) of the concentration-response curve for lethality in rats exposed to UF_6. An intraspecies species UF of 3 was applied and is considered sufficient because the steep concentration-response curve for lethality implies little intra-individual variability. A UF of 3 was also applied for interspecies variability (total UF = 10). Application of the full interspecies UF of 10 would result in AEGL-3 values inconsistent with the overall data set. For example, assuming the production of 4 mol of HF from the hydrolysis of 1 mol of UF_6, exposure at the proposed AEGL-3 values would be equivalent to exposure to HF at 60 ppm for 10 min, 20 ppm for 30 min, 10 ppm for 1 h, 2.5 ppm for 4 h, and 1.3 ppm for 8 h. No effects on respiratory parameters were noted in healthy humans after exposure to HF at 6.4 ppm for 1 h or after repeated exposures of up to 8.1 ppm (Lund et al. 1999; Largent 1960, 1961). Although this scenario does not take into account the uranium portion of the exposure, given the steepness of the concentration-response curve, it is unlikely that one would experience life-threatening effects at these concentrations. Therefore, an interspecies UF of 3 is justified. The value was then scaled to the 10-min, 30-min, 4-h, and 8-h time points using $C^1 \times t = k$. An exponent value of $n = 0.66$ was derived from rat lethality data from experiments ranging in duration from 2 min to 1 h in the key study. The exponent was rounded to 1.0 for extrapolation because the data used to derive the exponent were limited (one study) and the derived n is below the range of a normal dose-response curve. The AEGL-3 values for

TABLE 5-14 AEGL-3 Values for Uranium Hexafluoride (mg/m^3)

10 min	30 min	1 h	4 h	8 h
216	72	36	9.0	4.5

UF_6 are presented in Table 5-14 (above), and the calculations for those AEGL-3 values are presented in Appendix A. The relationship between AEGL-3 and animal exposures is shown in Figure 5-2.

8. SUMMARY OF PROPOSED AEGLS

8.1. AEGL Values and Toxicity End Points

The derived AEGL values for various levels of effects and durations of exposure are summarized in Table 5-15. Because no data consistent with the definition of AEGL-1 were available for the title compound, it was necessary to modify HF AEGL-1 values to approximate AEGL-1 values for UF_6. Renal effects in dogs were used as the basis for AEGL-2. An estimate of the concentration causing no deaths in rats was used for the AEGL-3.

8.2. Other Exposure Criteria

The established exposure criteria for UF_6 or soluble uranium compounds are presented in Table 5-16.

FIGURE 5-15 Summary of Proposed AEGLs (mg/m^3)

Classification	10 min	30 min	1 h	4 h	8 h
AEGL-1 (Nondisabling)	3.6	3.6	3.6	NR	NR
AEGL-2 (Disabling)	28	19	9.6	2.4	1.2
AEGL-3 (Lethal)	216	72	36	9.0	4.5

Abbreviation: NR, not recommended.

TABLE 5-16 Extant Standards and Guidelines for Uranium Hexafluoride (mg/m^3)

	Exposure Duration				
Guideline	10 min	30 min	1 h	4 h	8 h
AEGL-1	3.6	3.6	3.6	NR	NR
AEGL-2	28	19	9.6	2.4	1.2
AEGL-3	216	72	36	9.0	4.5
ERPG-1[a]	5				
ERPG-2[a]	15				
ERPG-3[a]	30				
NIOSH IDLH[b]	10 (soluble compounds as uranium) (UF_6 at 14.8 mg/m^3)				
NIOSH TWA[c]				0.05 (soluble compounds as uranium) (UF_6 at 0.07 mg/m^3)	
OSHA TWA[d]				0.05 (soluble compounds as uranium) (UF_6 at 0.07 mg/m^3)	
ACGIH TLV-TWA[e]				0.2 (soluble compounds as uranium) (UF_6 at 0.30 mg/m^3)	
ACGIH TLV-STEL[f]	0.6 (soluble compounds as uranium) (UF_6 at 0.89 mg/m^3)				

[a]ERPG (emergency response planning guidelines of the American Industrial Hygiene Association) (AIHA 1991). The ERPG-1 is the maximum airborne concentration below which it is believed nearly all individuals could be exposed for up to 1 h without experiencing symptoms other than mild, transient adverse health effects or without perceiving a clearly defined objectionable odor. The ERPG-1 for uranium hexafluoride is based on an estimated no-effect exposure level and irritation on conversion to HF. The ERPG-2 is the maximum airborne concentration below which it is believed nearly all individuals could be exposed for up to 1 h without experiencing or developing irreversible or other serious health effects or symptoms that could impair an individual's ability to take protective action. The ERPG-2 for uranium hexafluoride is based on renal injury. The ERPG-3 is the maximum airborne concentration below which it is believed nearly all individuals could be exposed for up to 1 h without experiencing or developing life-threatening health effects. The ERPG-3 for uranium hexafluoride is based on lethality data.

[b]IDLH (immediately dangerous to life and health standard of the National Institute of Occupational Safety and Health) (NIOSH 1997). The IDLH represents the maximum concentration from which one could escape within 30 min without any escape-impairing symptoms or any irreversible health effects. The IDLH for uranium hexafluoride is based on chronic toxicity data in animals.
[c]NIOSH-TWA (time-weighted average) (NIOSH 1997). The TWA is defined analogous to the ACGIH TWA.
[d]OSHA PEL-TWA (permissible exposure limits-time-weighted average of the Occupational Health and Safety Administration) (29 CFR § 1910.1000). The PEL-TWA is defined analogous to the ACGIH TLV-TWA but is for exposures of no more than 10 h/d, 40 h/wk.
[e]ACGIH TLV-TWA (Threshold Limit Value-time-weighted average of the American Conference of Governmental Industrial Hygienists) (ACGIH 1999). The TLV-TWA is the time-weighted average concentration for a normal 8-h work day and a 40-h work week to which nearly all workers may be repeatedly exposed, day after day, without adverse effect.
[f]ACGIH TLV-STEL (Threshold Limit Value-short-term exposure limit) (ACGIH 1999). The TVL-STEL is for a 15 min exposure.

8.3. Data Quality and Research Needs

Data appropriate for AEGL-1 derivation were not available, necessitating the use of HF AEGL-1 values to approximate AEGL-1 values for UF_6. Data appropriate for derivation of AEGL-2 values were limited to one study using two dogs. Data appropriate for derivation of AEGL-3 values were limited to one rat LC_{50} study with little reported experimental detail.

9. REFERENCES

ACGIH (American Conference of Governmental Industrial Hygienists, Inc). 1999. Threshold Limit Values (TLVs) for Chemical Substances and Physical Agents and Biological Exposure Indices (BEIs). Cincinnati, OH: ACGIH.
AIHA (American Industrial Hygiene Association). 1995. Emergency Response Planning Guidelines for Uranium Hexafluoride. Fairfax, VA: AIHA.
ATSDR (Agency for Toxic Substances and Disease Registry). 1993. Toxicological Profile for Fluorides, Hydrogen Fluoride, and Fluorines. U.S. Department of Health and Human Services, Washington, DC. April 1993.
ATSDR (Agency for Toxic Substances and Disease Registry). 1999. Toxicological Profile for Uranium. U.S. Department of Health and Human Services, Washington, DC. September 1999.

EPA (U.S. Environmental Protection Agency). 2001. Acute exposure guideline levels for hydrogen fluoride [interim draft 3, 7/2001].

Fisher, D.R., K.L. Kathren, and M.J. Swint. 1991. Modified biokinetic model for uranium from analysis of acute exposure to uranium hexafluoride. Health Phys. 60:335-342.

Just, R.A. 1984. Report on Toxicological Studies Concerning Exposures to UF_6 and UF_6 Hydrolysis Products. Oak Ridge Gaseous Diffusion Plant, operated by Martin Marietta Energy Systems, Inc., for the U.S. Department of Energy, Washington, DC.

Just, R.A., and V.S. Elmer. 1984. Generic Report on Health Effects for the U.S. Gaseous Diffusion Plants. Oak Ridge Gaseous Diffusion Plant, Oak Ridge, TN, operated by Martim Marietta Energy Systems, Inc., for the U.S. Department of Energy, Washington, DC.

Kathren, R.L. and R.H. Moore. 1986. Acute accidental inhalation of U: A 38-year follow-up. Health Phys. 51:609-619.

Largent, E.J. 1960. The metabolism of fluorides in man. AMA Arch. Ind. Health 21:318-323.

Largent, E.J. 1961. Fluorosis: The Health Aspects of Fluorine Compounds. Columbus, OH: Ohio State University Press. Pp. 34-39, 43-48.

Leach, L.J., R.M. Gelein, B.J. Panner, C.L. Yulie, C.C. Cox, M.M. Balys, and P.M. Rolchigo. 1984. Acute Toxicity of the Hydrolysis Products of Uranium Hexafluoride (UF_6) when Inhaled by the Rat and Guinea Pig. Final Report (K/SUB/81-9039/3). Rochester, NY: University of Rochester Medical Center.

Lund, K., M. Refsnes, T. Sandstrom, P. Sostrand, P. Schwarze, J. Boe, and J. Kongerud. 1999. Increased CD3 positive cells in bronchoalveolar lavage fluid after hydrogen fluoride inhalation. Scand. J. Work Environ. Health 25:326-334.

Machle, W., and K. Kitzmiller. 1935. The effects of the inhalation of hydrogen fluoride: II. The response following exposure to low concentrations. J. Ind. Hyg. 17:223-229.

Machle, W., F. Thamann, K. Kitzmiller, et al. 1934. The effects of the inhalation of hydrogen fluoride: I. The response following exposure to high concentrations. J. Ind. Hyg. 16:129-145.

Moore, R.H., and R.L. Kathren. 1985. A World War II uranium hexafluoride inhalation event with pulmonary implications for today. J. Occup. Med. 27:753-756.

Morrow, P., R. Gelein, H. Beiter, J. Scott, J. Picano, and C. Yulie. 1982. Inhalation and intraveneous studies of UF_6 and UO_2F_2 in dogs. Health Phys. 43:859-873.

NIOSH (National Institute of Occupational Safety and Health). 1997. NIOSH Pocket Guide to Chemical Hazards. U.S. Department of Health and Human Services, Washington, DC.

Nuclear Regulatory Commission. 1986. Assessment of the Public Health Impact from the Accidental Release of UF_6 at the Sequoyah Fuels Corporation Facility at Gore, Oklahoma. NUREG-1189, Vols. 1, 2. U.S. Nuclear Regulatory Commission, Rockville, MD.

Spiegel, C.J. 1949. Uranium Hexafluoride. Pp. 532-547 in Pharmacology and Toxicology of Uranium Compounds. New York, NY: McGraw-Hill Book Company.

Stokinger, H.E. 1949. Toxicity following inhalation of fluorine and hydrogen fluoride. Pp. 1021-1057 in Pharmacology and Toxicology of Uranium Compounds, C. Voegtlin and H.C. Hodge, eds. New York, NY: McGraw Hill.

ten Berge, W.F., A. Zwart, and L.M. Appelman. 1986. Concentration-time mortality response relationship of irritant and systemically acting vapours and gases. J. Hazard. Mat. 13:301-309.

UNSCEAR (United Nations Scientific Committee on the Effects of Atomic Radiation). 1993. Sources and Effects of Ionizing Radiation. New York: United Nations.

APPENDIX A

Derivation of AEGL Values

Derivation of AEGL-1 Values

Key study: Based on HF AEGL-1 values (Lund et al. 1999)

10-min, 30-min, and 1-h HF AEGL-1: 1.0 ppm

Stoichiometric adjustment factor: 4, because 4 mol of HF are produced from hydrolysis of 1 mol of UF_6.

10-min, 30-min, and 1-h AEGL-1: AEGL = 1.0 ppm ÷ 4 = 0.25 ppm;
AEGL = 0.25 ppm × 14.4 = 3.6 mg/m³

Derivation of AEGL-2 Values

Key study: Morrow et al., 1982

Toxicity end points: Renal pathology in dogs.

Time-scaling: *10-min and 30-min*

$C^3 \times t = k$

$(192 \text{ mg/m}^3)^3 \times 0.5 \text{ h} = 3{,}538{,}944 \text{ mg/m}^3\cdot\text{h}$

1-h, 4-h, and 8-h

$C^1 \times t = k$

$(192 \text{ mg/m}^3)^1 \times 0.5 \text{ h} = 96 \text{ mg/m}^3\cdot\text{h}$

Uncertainty factors: 3 for interspecies variability
3 for intraspecies variability

10-min AEGL-2: $C^3 \times 0.167\ \text{h} = 3{,}538{,}944\ \text{mg/m}^3\cdot\text{h}$
$C^3 = 21191281\ \text{mg/m}^3$
$C = 276.7\ \text{mg/m}^3$
10-min AEGL-2 = 276.7/10 = 27.7 mg/m^3

30-min AEGL-2: $C^3 \times 0.5\ \text{h} = 3{,}538{,}944\ \text{mg/m}^3\cdot\text{h}$
$C^3 = 7{,}077{,}888\ \text{mg/m}^3$
$C = 192\ \text{mg/m}^3$
30-min AEGL-2 = 192/10 = 19.2 mg/m^3

1-h AEGL-2: $C^1 \times 1\ \text{h} = 96\ \text{mg/m}^3\cdot\text{h}$
$C^1 = 96\ \text{mg/m}^3$
$C = 96\ \text{mg/m}^3$
1-h AEGL-2 = 96/10 = 9.6 mg/m^3

4-h AEGL-2: $C^1 \times 4\ \text{h} = 96\ \text{mg/m}^3\cdot\text{h}$
$C^1 = 24\ \text{mg/m}^3$
$C = 24\ \text{mg/m}^3$
4-h AEGL-2 = 24/10 = 2.4 mg/m^3

8-h AEGL-2: $C^1 \times 8\ \text{h} = 96\ \text{mg/m}^3\cdot\text{h}$
$C^1 = 12\ \text{mg/m}^3$
$C = 12\ \text{mg/m}^3$
8-h AEGL-2 = 12/10 = 1.2 mg/m^3

Derivation of AEGL-3 Values

Key study: Leach et al. 1984

Toxicity end point: Estimated 1-h threshold for death in rats (one-third the LC_{50})

Time-scaling: $C^1 \times t = k$
$(365)1 \times 1\ \text{h} = 365\ \text{mg/m}^3\cdot\text{h}$

Uncertainty factors: 3 for interspecies variability
3 for intraspecies variability

10-min AEGL-3: $C^1 \times 0.167\ h = 365\ mg/m^3{\cdot}h$
$C = 2{,}165.6\ mg/m^3$
10-min AEGL-3 = 2,165.6/10 = 216 mg/m^3

30-min AEGL-3: $C^1 \times 0.5\ h = 365\ mg/m^3{\cdot}h$
$C = 730\ mg/m^3$
30-min AEGL-3 = 730/10 = 73 mg/m^3

1-h AEGL-3: $C^1 \times 1\ h = 365\ mg/m^3{\cdot}h$
$C^1 = 49\ mg/m^3{\cdot}h$
$C = 364\ mg/m^3$
1-h AEGL-3 =364/10 = 36.4 mg/m^3

4-h AEGL-3: $C^1 \times 4\ h = 365\ mg/m^3{\cdot}h$
$C = 90\ mg/m^3$
4-h AEGL-3 = 90/10 = 9.0 mg/m^3

8-h AEGL-3: $C^1 \times 8\ h = 365\ mg/m^3{\cdot}h$
$C = 45.6\ mg/m^3$
8-h AEGL-3 = 45.6/10 = 4.5 mg/m^3

APPENDIX B

ACUTE EXPOSURE GUIDELINE LEVELS FOR URANIUM HEXAFLUORIDE (CAS No. 7783-81-5)

DERIVATION SUMMARY

AEGL-1				
10 min	30 min	1 h	4 h	8 h
3.6 mg/m^3	3.6 mg/m^3	3.6 mg/m^3	NR	NR
Key Reference: EPA (U.S. Environmental Protection Agency). 2001. Acute exposure guideline levels for hydrogen fluoride [interim draft 3, 7/2001]. (Lund et al. 1999. Increased CD3 positive cells in bronchoalveolar lavage fluid after hydrogen fluoride inhalation. Scand. J. Work. Environ. Health 25:326-334.)				
Test species/strain/number: 20 healthy male volunteers				
Exposure route/concentrations/durations: Inhalation, average concentrations at 0.2-0.7 ppm (n = 9), 0.85-2.9 ppm (n = 7), and 3.0-9.3 ppm (n = 7).				
Effects: 0.2-0.7 ppm: No to low sensory irritation; no change in FVC, FEV_1; no inflammatory response in bronchoalveolar lavage fluid (BAL). 0.85-2.9 ppm: No to low sensory irritation; no change in FVC, FEV_1; increase in the percentage of CD3 cells and myeloperoxidase in bronchial portion of BAL; no increases in neutrophils, eosinophils, protein, or methyl histamine in BAL. 3.0-6.3 ppm: Low sensory irritation; no change in FVC, FEV_1; increase in the percentage of CD3 cells and myeloperoxidase in bronchial portion of BAL; no increases in neutrophils, eosinophils, protein, or methyl histamine in BAL.				
End point/concentration/rationale: The subthreshold concentration for inflammation at 3 ppm (0.85-2.9 ppm) for 1 h, which was without sensory irritation, was chosen as the basis for the AEGL-1.				
Uncertainty factors/rationale: Total uncertainty factor: 3 Interspecies: Not applicable since human subjects were the test species. Intraspecies: 3. The subjects were healthy adult males. The concentration used is far below tested concentrations that did not cause symptoms of bronchial constriction in healthy adults (ranges up to 6.3 ppm [Lund et al. 1997] and 8.1 ppm [Largent 1960, 1961]).				

AEGL-1 *Continued*
Stoichiometric adjustment factor: 4. A maximum of 4 mol of HF may be produced by hydrolysis from 1 mol of UF_6
Animal to human dosimetric adjustment: Not applicable, human data used
Time-scaling: Not applied. AEGL-1 values for HF were calculated by adjusting the 1-h concentration of 3 ppm by an UF of 3. Because the response would be similar at shorter exposure durations, the 10- and 30-minute values were set equal to the 1-h concentration. The UF_6 AEGL-1 values were derived by applying a modifying factor of 4 to the HF AEGL values, because a maximum of 4 mol of HF are produced for every mole of UF_6 hydrolyzed. AEGL-1 values for UF_6 were derived only for the 10-min, 30-min, and 1-h time points, because derivation of 4- and 8-h values results in AEGL-1 values greater than the 4- and 8-h AEGL-2s, which would be inconsistent with the total database.
Data quality: Data appropriate for AEGL-1 derivation were not available for UF_6. Therefore, AEGL-1 values for UF_6 were approximated from the AEGL-1 values for HF, a known hydrolysis product and a likely source of respiratory irritation. For each mole of UF_6, 4 mol of HF may be produced by hydrolysis. Thus, the HF AEGL-1 was divided by a factor of 4 to approximate an AEGL-1 for UF_6.

AEGL-2				
10 min	30 min	1 h	4 h	8 h
28 mg/m^3	19 mg/m^3	9.6 mg/m^3	2.4 mg/m^3	1.2 mg/m^3
Key reference: Morrow et al. 1982. Inhalation and intravenous studies of UF_6 and UO_2F_2 in dogs. Health Physics. 43: 859-873.				
Test species/strain/number: 1 female beagle dog per concentration				
Exposure route/concentration/durations: Inhalation, 192-284 mg/m^3 for 30 min to 1 h.				
Effects: At 192 mg/m^3, kidney pathology; at 284 mg/m^3, kidney pathology				
End point/concentration/rationale: 192 mg/m^3 for 30 min, LOEL for kidney pathology				
Uncertainty factors/rationale: Total uncertainty factor: 10 This total UF is considered sufficient because the use of a larger total UF would yield AEGL-2 values below or approaching the AEGL-1 values, which are considered to be no-observed-effect levels and were based on a threshold for inflammation in humans (HF study). Also, humans were exposed to HF repeatedly at up to 8 ppm with only slight nasal irritation; that concentration is stoichiometrically equivalent to UF_6 at 28.8 mg/m^3. Interspecies: 3 Intraspecies: 3				
Modifying factor: NA				
Animal to human dosimetric adjustment: None, insufficient data				
Time-scaling: To obtain conservative and protective AEGL values in the absence of an empirically derived, chemical-specific scaling exponent, temporal scaling was performed using the $C^n \times t = k$ equation where $n = 3$ when extrapolating to shorter time points and $n = 1$ when extrapolating to longer time points. (Although a chemical-specific exponent of 0.66 was derived from rat lethality data, the default values were utilized for time-scaling AEGL-2 values because the end points for AEGL-2 [renal toxicity] and for death [pulmonary edema] involve different mechanisms of action).				
Data quality: Although only one dog was exposed at each treatment level, the use of that data is supported by the fact that Morrow et al. (1982) similarly exposed 17 dogs to UO_2F_2 at 200-270 mg/m^3 for 30 min to 2.5 h and observed similar renal pathology. Although the UO_2F_2 is not the title compound, it is the hydrolysis product of UF_6 responsible for the renal effects.				

AEGL-3				
10 min	30 min	1 h	4 h	8 h
216 mg/m³	72 mg/m³	36 mg/m³	9.0 mg/m³	4.5 mg/m³
Key reference: Leach et al. 1984. Acute toxicity of the hydrolysis products of uranium hexafluoride (UF_6) when inhaled by the rat and guinea pig. Final Report. (K/SUB/81-9039/3). University of Rochester Medical Center. Rochester, NY.				
Test species/strain/gender/number: Long Evans rats, males				
Exposure route/concentrations/durations: Inhalation, 1 h				
Effects: LC_{50}, 1,095 mg/m³				
End point/concentration/rationale: Estimated 1-h threshold for death (one-third LC_{50}), 365 mg/m³				
Uncertainty factors/rationale: Total uncertainty factor: 10 Interspecies: 3. An application of the full interspecies UF of 10 would result in AEGL-3 values that are inconsistent with the overall data set. For example, assuming the production of 4 mol of HF from the hydrolysis of 1 mol of UF_6, exposure at the proposed AEGL-3 values would be equivalent to exposure to HF at 60 ppm for 10 min, 20 ppm for 30 min, 10 ppm for 1 h, 2.5 ppm for 4 h, and 1.3 ppm for 8 h. No effects on respiratory parameters were noted in healthy humans after exposure to HF at 6.4 ppm for 1 h or after repeated exposures at up to 8.1 ppm (Lund et al. 1999; Largent 1960, 1961). Although this scenario does not take into account the uranium portion of the exposure, given the steepness of the concentration-response curve, it is unlikely that one would experience life-threatening effects at these concentrations. Intraspecies: 3. The steep concentration-response curve implies limited intra-individual variability.				
Modifying factor: Not applicable				
Animal to human dosimetric adjustment: Insufficient data				
Time-scaling: $C^n \times t = k$ where $n = 1$ on the basis of regression analysis of rat lethality data from experiments ranging in duration from 2 min to 1 h. Calculated value of $n = 0.66$ was rounded to $n = 1$ because limited data were used to derive n and the calculated value is below the range of a normal dose-response curve (Leach et al. 1984).				

APPENDIX C

Time-Scaling Calculation and Key Study Description for Hydrogen Fluoride AEGL-1 (EPA 2001)

Key study:	Lund et al. 1997, 1999
Toxicity end point:	Biomarkers of exposure during 1-h exposure of human subjects at several ranges of concentrations.
Time-scaling:	Not applied
Uncertainty factors:	3 for differences in human sensitivity (effects are unlikely to differ among individuals). This concentration should be protective of asthmatic individuals because it is below average (2 ppm) and ranges of concentrations (up to 8.1 ppm) (Largent 1960, 1961) that produced slight to mild irritation in healthy adult male subjects.
Calculations:	The 3 ppm concentration was divided by an intraspecies UF of 3. The resulting concentration, 1 ppm, was used for the 10- and 30-min and 1-h, 4-h, and 8-h AEGL-1.

Summary of Key Study and Rationale Used in the Hydrogen Fluoride AEGL-1 Derivation (EPA 2001)

The AEGL-1 was based on a concentration of 3 ppm (range, 0.85-2.93 ppm) for 1 h, which was the threshold for pulmonary inflammation as evidenced by an increase in the percentages of several inflammatory parameters, such as CD3 cells and myeloperoxidase, in the bronchoalveolar lavage fluid of 20 healthy adult human subjects (Lund et al. 1999). There were no increases in neutrophils, eosinophils, protein, or methyl histamine at this or the next higher average exposure concentration of 4.7 ppm (range, 3.05-

6.34 ppm). There were no changes in lung function and no to minor increased symptoms of irritation at this concentration (Lund et al. 1997). Although healthy adults were tested, several individuals had increased immune factors, indicating atopy. The 3 ppm concentration was divided by an intraspecies UF of 3 because the response to an irritant gas is not expected to differ more than 3-fold among individuals, including susceptible individuals such as asthmatic people. The 1 ppm concentration was used for the 10-min, 30-min, and 1-h exposure durations. The 1 ppm concentration was reduced by a factor of 2 for the 4- and 8-h exposure durations. Because there were no effects on respiratory parameters of healthy adults at concentrations that ranged up to 6.34 ppm in this study and up to 8.1 ppm with repeated exposures in a supporting study (Largent 1960, 1961), the calculated AEGL-1 values will probably be protective of the asthmatic population.